Daniel Georg Possner

Organic Peracid Etches

Daniel Georg Possner

Organic Peracid Etches

Novel chromium-free etching solutions for the delineation of crystalline defects in thin silicon films

Südwestdeutscher Verlag für Hochschulschriften

Impressum/Imprint (nur für Deutschland/ only for Germany)
Bibliografische Information der Deutschen Nationalbibliothek: Die Deutsche Nationalbibliothek verzeichnet diese Publikation in der Deutschen Nationalbibliografie; detaillierte bibliografische Daten sind im Internet über http://dnb.d-nb.de abrufbar.

Alle in diesem Buch genannten Marken und Produktnamen unterliegen warenzeichen-, marken- oder patentrechtlichem Schutz bzw. sind Warenzeichen oder eingetragene Warenzeichen der jeweiligen Inhaber. Die Wiedergabe von Marken, Produktnamen, Gebrauchsnamen, Handelsnamen, Warenbezeichnungen u.s.w. in diesem Werk berechtigt auch ohne besondere Kennzeichnung nicht zu der Annahme, dass solche Namen im Sinne der Warenzeichen- und Markenschutzgesetzgebung als frei zu betrachten wären und daher von jedermann benutzt werden dürften.

Verlag: Südwestdeutscher Verlag für Hochschulschriften GmbH & Co. KG
Dudweiler Landstr. 99, 66123 Saarbrücken, Deutschland
Telefon +49 681 37 20 271-1, Telefax +49 681 37 20 271-0
Email: info@svh-verlag.de
Zugl.: Frankfurt am Main, Goethe-Universität, Diss., 2010

Herstellung in Deutschland:
Schaltungsdienst Lange o.H.G., Berlin
Books on Demand GmbH, Norderstedt
Reha GmbH, Saarbrücken
Amazon Distribution GmbH, Leipzig
ISBN: 978-3-8381-1853-6

Imprint (only for USA, GB)
Bibliographic information published by the Deutsche Nationalbibliothek: The Deutsche Nationalbibliothek lists this publication in the Deutsche Nationalbibliografie; detailed bibliographic data are available in the Internet at http://dnb.d-nb.de.

Any brand names and product names mentioned in this book are subject to trademark, brand or patent protection and are trademarks or registered trademarks of their respective holders. The use of brand names, product names, common names, trade names, product descriptions etc. even without a particular marking in this works is in no way to be construed to mean that such names may be regarded as unrestricted in respect of trademark and brand protection legislation and could thus be used by anyone.

Publisher: Südwestdeutscher Verlag für Hochschulschriften GmbH & Co. KG
Dudweiler Landstr. 99, 66123 Saarbrücken, Germany
Phone +49 681 37 20 271-1, Fax +49 681 37 20 271-0
Email: info@svh-verlag.de

Printed in the U.S.A.
Printed in the U.K. by (see last page)
ISBN: 978-3-8381-1853-6

Copyright © 2010 by the author and Südwestdeutscher Verlag für Hochschulschriften GmbH & Co. KG and licensors
All rights reserved. Saarbrücken 2010

List of abbreviations

A: preexponential factor (Arrhenius equation)

AFM:. atomic force microscopy

BMD: bulk micro defects

BOX: buried oxide layer

c: concentration

COP: crystal originated particle (D-defect, void)

C_V: vacancy concentration

CVD: chemical vapour deposition

CZ: Czochralski grown material

D: drain (transistor)

D: diffusion coefficient

dd: defect density

dil.: diluted

DNA: deoxyribonucleic acid

DZ: denuded zone

E_a: activation energy

e. g.: exempli gratia (for example)

FGMOSFET: floating gate metal oxide semiconductor field-effect transistor

FZ: Floatzone material

G: gate (transistor)

Ge: Germanium

G: axial temperature gradient (crystal growth process)

GeOI: Germanium On Insulato

GOI: Gate Oxide Integrity

HAc: acetic acid

HD: highly doped

HF: hydrofluoric acid

I: current

IC: integrated circuit

IR: Infrared

K: equilibrium constant

k: rate coefficient

kWh: kilowatt hour(s)

LST: Laser Scattering Tomography

n: negative

MGS: metallurgical grade silicon

MOSFET: metal oxide semiconductor field-effect transistor

ppb: parts per Billion

ppm: parts per million

ppt: parts per Trillion

nm: nanometer

N: Newton

NMOS: n-doped metal oxide semiconductor field-effect transistor

OPE: Organic Peracid Etch

orig.: original

p: positive

PMOS: p-doped metal oxide semiconductor field-effect transistor

R: gas-law constant = 8.414 $JK^{-1}mol^{-1}$

r: removal rate

rpm: revolutions per minute

RTP: rapid thermal processing

S: Selectivity (of an etching solution)

S: Source (transistor)

SC1: Standard Cleaning 1

SC2: Standard Cleaning 2

SEM: scanning electron microscopy

Si: silicon

sSi: strained silicon

SiGe: silicon/germanium alloy

SIMOX: Separation by Implanted Oxygen

SOI: Silicon On Insulator

sSOI: strained Silicon On Insulator

T: temperature [°C or K]

t: time

TD: threading dislocation

TEM: transmission electron microscopy

U: voltage

V: pulling speed (crystal growth process)

vs: versus

ξ_T: critical ratio

η: viscosity

µl: microliter

µm: micrometer

σ: standard deviation

Ω: resistivity

Table of contents

List of abbreviations II
Table of contents V
List of tables VIII
List of illustrations XI
1 Introduction 1
2 Crystal growth process and wafer manufacturing 6
   2.1 Introduction 6
   2.2 Preparation of high-purity silicon 7
   2.3 Crystal growth methods 9
      2.3.1 The Czochralski Method 9
      2.3.2 The Float-Zone Method 11
   2.4 Wafer manufacturing 13
      2.4.1 Slicing and polishing 13
      2.4.2 Wafer cleaning 15
   2.5 Epitaxy 16
   2.6 New materials 19
      2.6.1 Introduction 19
      2.6.2 Silicon on Insulator (SOI) 19
      2.6.3 Strained Silicon on Insulator (sSOI) 22
3 Crystal defects 26
   3.1 Introduction 26
   3.2 Dislocations 28
   3.3 Stacking faults 30
   3.4 Point defects and their agglomerates 31
      3.4.1 Introduction 31
      3.4.2 Swirl defects (A- and B-defects) 31
      3.4.3 D-defects (vacancy agglomerates or COPs) 34
      3.4.4 Variation of V and G during the crystal growth process 35
      3.4.5 Oxygen precipitates 36
      3.4.6 Oxidation-induced stacking faults (OSF) and OSF ring 38
   3.5 Perfect (nearly defect free) silicon crystals 40
      3.5.1 Epitaxy-process: 40

3.5.2 Crystal growth process: 41
3.5.3 Argon annealing: 41
4 Delineation of crystal defects by chemical etching 42
 4.1 Introduction 42
 4.2 General aspects of etching silicon 44
  4.2.1 Diffusion-controlled etching mechanism 44
  4.2.2 Reaction-controlled etching mechanism 45
  4.2.3 Characterisation of the etching mechanism by its activation energy E_a 46
 4.3 Classification of etching solutions 46
  Polishing etches 46
  Structural etches 47
 4.4 Chromium based etch recipes 50
  4.4.1 The Secco Solution 50
  4.4.2 Further chromium based etch recipes 55
  4.4.3 Toxicity of chromium (VI) compounds 56
 4.5 Existing chromium free recipes 57
 4.6 Chemistry of HNO_3/HF mixtures 59
  Etching mechanism of silicon in HF/HNO_3 systems 61
5 Organic Peracid Etches: A new class of chromium free etching solutions for the delineation of defects in different silicon based materials 64
 5.1 Introduction 64
 5.2 The H_2O_2/HF/HAc system 66
  5.2.1 Introduction 66
  5.2.2 Experimental procedure 67
  5.2.3 Influence of the hydrofluoric acid content on etching behaviour 68
  5.2.4 Influence of the concentration of hydrogen peroxide and peracetic acid on removal rate 68
  5.2.5 Peracetic acid content as a function of time 69
  5.2.6 Dependence of removal rate on PAA content 74
  5.2.7 Delineation of crystal defects 75
 5.3 Etching solutions containing perpropanoic acid 76
 5.4 Etching solutions containing perbutyric acid 79
 5.5 Etching solutions containing performic acid 80
 5.6 Comparison of the different etch recipes 80

5.6.2 Removal vs. etching time	81
5.6.3 Influence of temperature	82
5.6.4 Influence of stirring	83
5.7 Experimental results on epitaxial silicon wafers	83
5.8 Experimental results on silicon substrates	87
5.9 Experimental results on SOI material	90
5.9.2 Influence of the hydrofluoric acid content on the size of the etched figures	92
5.9.3 OPE tested on standard Smart-Cut^{TM} material	94
5.9.4 Experimental results on thin SOI	103
5.9.5 Experimental results on Simox material	108
5.9.6 Depth distribution of crystal defects in the SOI layer	111
5.10 Detailed characterisation of the etched figures by atomic force microscopy (AFM), scanning electron microscopy (SEM) and transmission electron microscopy (TEM)	115
5.10.1 AFM investigations	115
5.10.2 Characterisation of the etch hillocks by SEM and TEM	120
5.11 Experimental results on sSOI material	124
6 Characterization of the etching solutions on the basis of their selectivity, activation energy and standard potential	131
6.1 Introduction	131
6.2 Experimental determination of the selectivity	131
6.3 Calculation of the activation energy for the etching process at a crystal imperfection (dislocation)	144
6.4 Experimental determination of the standard potentials	145
6.5 Comparison of the different etching solutions by their selectivities, activation energies for the etching process and normal potentials	147
7.1 Summary and Conclusion	151
8.1 Outlook	155
7.2 Zusammenfassung	156
8.2 Ausblick	160
9 Appendix	161
9.1 Bibliography	161
9.2 Instruments used	165
9.3 Chemicals used:	166

List of tables

Table 1: Summary of the stages in wafer manufacturing. _____ 6
Table 2: Stages in the cleaning of silicon wafers. _____ 16
Table 3: Different silanes which can be used for the epitaxy process. _____ 17
Table 4: Different crystal defects _____ 27
Table 5: Morphology of the oxygen precipitates found by weak-beam TEM investigations. 38
Table 6: Compositions and properties of the different Secco recipes. The removal rates were determined on silicon substrates and SOI material.. _____ 53
Table 7: Experimentally determined activation energies of the different Secco solutions. _____ 55
Table 8: Compositions of further currently used chromium based etching solutions. _____ 56
Table 9: Chromium free etching solutionsbased on the HF/HNO3/HAc system _____ 57
Table 10: Experimentally determined activation energies of the MEMC and Jeita solution. 59
Table 11: Properties of the original and the modified Dash solution. _____ 65
Table 12: Composition of organic peracid etches with varying HF concentrations and the removal rates. _____ 68
Table 13: Effect on etching behaviour of replacing the alkanoic acid with water _____ 69
Table 14: Properties of different H_2O_2/HAc/HF mixtures. The maximum PAA content is obtained after 8-72 h, dependent on the composition of the etching solution. _____ 71
Table 15: Composition and properties of different Organic Peracid Etches. The removal rate increases with increasing PAA content. _____ 74
Table 16: Properties of the different OPE mixtures containing peracetic and perpropanoic acid. All equilibrium constants K were determined at 21°C. _____ 77
Table 17: Composition and properties of Organic Peracid Etches containing peracetic, perpropanoic and perbutyric acid as oxidizing agents. All equilibrium constants were determined at 21°C. _____ 80
Table 18: Organic Peracid Etches tested on different materials. _____ 81
Table 19: Experimentally determined activation energies for the etching process of silicon. 83
Table 20: Defect densities obtained on epitaxial Si with different OPE mixtures and the Secco diluted 1 as reference. _____ 85
Table 21: Physical properties of the silicon substrates used. _____ 88
Table 22: Defect densities found after etching with the OPE C and the Secco reference in different silicon substrates. The removal was calculated. _____ 90

List of tables

Table 23: Etching solutions with varying HF concentrations and the diameters of the corresponding etched figures. 93

Table 24: Defect densities found after etching with the Secco diluted 1, OPE C and OPE F. 96

Table 25: Defect densities obtained on three different OPE mixtures tested on standard Smart-Cut^{TM} material with a high OSF density. 101

Table 26: The OPE mixtures C, D and F were also tested on standard Smart-Cut^{TM} material with a low OSF density. The OSF density found after etching with the OPE C is 2 times higher than those found after etching with the Secco diluted reference and the OPE F. 102

Table 27: Defect densities in thin SOI found after etching with the different OPE mixtures and the Secco diluted 1 reference. All values are mean values. 105

Table 28: Comparison of the Swirl-defect densities found after etching with OPE A, D and the Secco diluted reference. 107

Table 29 Table 28: Comparison of defect densities found after etching in SOI material produced by the SIMOX technique. All values are mean values. 110

Table 30: Defect densities obtained after different etching times. All values are mean values. 114

Table 31: Properties of the different SOI materials. 118

Table 32: Summary of the experimental results of Smart-Cut^{TM} SOI materials etched with the OPE D. Etch hillocks were found in all cases 119

Table 33: Summary of the experimental results. The Organic Peracid Etches produce both pits and hillocks. 120

Table 34: Defect densities found after etching with the OPE A, B and F and the Secco diluted 2 reference. 129

Table 35: Physical and chemical properties of the Si-Bulk material used for the experiments 132

Table 36: Characterization of the etch pits found after treatment with different structural etches. 134

Table 37: Experimentally determined removal rates of the different etch formulations. 140

Table 38: Depth and diameter of the etch pits and selectivity (S) determined for the Secco solution at different etching times. 140

Table 39: Dependence of the selectivity on etching time determined for the Secco diluted 1 and Secco diluted 2 solution. The selectivity also decreases with increasing removal. 141

Table 40: Experimentally determined selectivities for the OPE C, D and F. 142

Table 41: Experimentally determined selectivities for the Jeita and MEMC solution. 143

List of tables

Table 42: Removal rates determined at defect-free sites and also at dislocations and selectivities of the different etching solutions discussed before. _____ 143

Table 43: Experimentally determined and calculated activation energies for the etching process of silicon. _____ 145

Table 44: Experimentally determined standard potentials. _____ 146

Table 45: A comparison of etching solutions on the basis of various experimentally determined parameters. _____ 147

List of illustrations

Figure 1: Construction of an n-doped metal oxide semiconductor field-effect transistor (NMOS). _____ 2

Figure 2: Operating NMOS (UGS > Uth). A conducting channel is formed between source and drain. A current flows from source to drain. _____ 3

Figure 3: Structure of a floating gate metal oxide semiconductor field-effect transistor (FGMOSFET). The floating gate is separated from the control gate by a thin silicon dioxide layer _____ 3

Figure 4: Formation of $SiHCl_3$ followed by fractional distillation. _____ 8

Figure 5: The $SiHCl_3$ is decomposed to polycrystalline silicon at the heated U–shaped silicon rods. _____ 8

Figure 6: A typical Czochralski silicon crystal growing system. _____ 10

Figure 7: A Czochralski grown silicon crystal. _____ 11

Figure 8: The Float-Zone process. _____ 13

Figure 9: Slicing of wafers with an annular or internal diameter saw. _____ 14

Figure 10: Slicing with a multiple wire saw. _____ 14

Figure 11: Formation of a hydrophobic surface after HF treatment and elimination of water. 16

Figure 12: Epitaxial layer grown on a substrate wafer by chemical vapour deposition (CVD). _____ 17

Figure 13: Vertical epitaxy reactor. _____ 17

Figure 14: Structure of SOI wafer. _____ 19

Figure 15: The SIMOX-process. _____ 20

Figure 16: Smart-Cut process for SOI production. _____ 21

Figure 17: High resolution TEM image of the Smart-Cut region just before splitting. _____ 21

Figure 18: Hydrophilic interactions. _____ 22

Figure 19: Formation of chemical bonds after annealing and elimination of water. _____ 22

Figure 20: Structure of sSOI _____ 22

Figure 21: Strained silicon on a virtual substrate. _____ 23

Figure 22: Crystal lattice of strained silicon. _____ 23

Figure 23: Raman spectrum of fabricated sSOI. _____ 24

Figure 24: sSOI fabrication. After bonding the virtual substrate is removed by ion implantation followed by annealing and selectively etching. _____ 24

List of illustrations

Figure 25: Simple two-dimensional model of a crystal lattice showing different types of crystal defects. _____ 28

Figure 26: An extra half plane of atoms is added to the crystal lattice forming an edge dislocation. _____ 29

Figure 27: A screw dislocation. _____ 29

Figure 28: Illustration of Burger´s circuit and Burger´s vector. _____ 30

Figure 29: Single dislocation loops found below the "Swirl hillocks" after etching. _____ 32

Figure 30: Simple arrangement of dislocation loops. It is possible to distinguish the individual loops. _____ 32

Figure 31: Complex arrangement of dislocation loops. It is not possible to distinguish between the individual loops. _____ 33

Figure 32: HVEM images of decorated dislocation loops. _____ 33

Figure 33: Cross sectional TEM image of a D-defect. The octahedral twin structure can be recognized. _____ 34

Figure 34: Oxygen interstitial atom (Oi) incorporated in the crystal lattice. _____ 36

Figure 35: Oxidation-induced stacking faults, revealed by Secco etching. _____ 39

Figure 36: Surface OSF on CZ material. Found after etching with Secco solution (3 min.). _____ 39

Figure 37: X-ray topograph of an OSF ring. The CZ wafer was first heated to 1200°C in a wet O_2 atmosphere and then etched. _____ 40

Figure 38: "Nearly defect free" epitaxial layer grown on a substrate wafer. _____ 40

Figure 39: Equal attack of a polishing etch at a crystal surface. _____ 47

Figure 40: Activation energy for the etching process at a perfect crystal lattice. EA = activation energy for etching process. A = initial state, B = final state _____ 47

Figure 41: Schematic representation of the difference in activation energies between a perfect lattice, and a defect site. A^P = initial state (perfect crystal lattice) _____ 48

Figure 42: The removal rates of a non-preferential etching solution. _____ 48

Figure 43: The removal rates of a preferential etch (C1 - C4). _____ 49

Figure 44: Possible reaction mechanism for the reduction of chromium(VI)-compounds in acid solutions. The first step is the adsorption of chromate at the silicon surface. The next step is an electron transfer from the silicon to the chromium(VI)-species. _____ 52

Figure 45: Removal rate vs temperature determined on standard Si-Bulk material. _____ 53

Figure 46: Temperature dependence of the removal rates shown for the Secco-diluted 1 and the Secco-diluted 2 solutions. The removal rates were estimated on SOI material. _____ 54

List of illustrations

Figure 47: The activation energies of the original Secco recipe and the Secco diluted 1 and Secco diluted 2 solution. r = removal rate and T= temperature. _____ 54

Figure 48: Removal rate as a function of temperature. _____ 58

Figure 49: Etching mechanism of silicon in HF/HNO_3 mixtures after Steinert et al. The first step is the partial oxidation of silicon which leads to the formation of holes (h+). _____ 62

Figure 50: The second step is a nucleophilic attack of F- ions which results in the formation of Si-F bonds. _____ 62

Figure 51: The polarized Si-Si bonds are attacked by HF. _____ 62

Figure 52: The last step is the formation of $SiF4$ _____ 63

Figure 53: PAA content as a function of time determined for OPE A. _____ 72

Figure 54: PAA content of OPE B as a function of time. _____ 73

Figure 55: PAA content of OPE C as a function of time. _____ 73

Figure 56: The removal rate as a function of PAA content. _____ 75

Figure 57: Optical micrograph of a CZ Si fragment after scratching and heating in a copper contaminated furnace.This fragment was etched with OPE C for 24 h at room temperature. Different kinds of defects can be recognized: dislocations (black oval shaped etch pits), vacancy agglomerates (square shaped pits) and copper decorated D-defects (small hillocks). _____ 76

Figure 58: SEM image of a CZ Si wafer fragment after etching with OPE D. _____ 78

Figure 59: Wafer fragment of CZ-grown Si scratched and heated to generate dislocations. The black oval-shaped pits found after etching with OPE D (etching time: 24 h, removal: ~ 2.5 µm) are caused by dislocations. _____ 78

Figure 60: Swirl defects found after etching with OPE D in thin SOI material. The Swirl-defects were delineated in a small zone of silicon bulk at the edge of the SOI wafer. Etching time: 6 minutes, removal: ~ 10 nm. _____ 79

Figure 61: Removal as a function of etching time for different Organic Peracid Etches. _____ 82

Figure 62: Exponential dependence of the removal rate on temperature. _____ 82

Figure 63: Nearly defect-free epitaxial layer _____ 84

Figure 64: Wafer fragment etched with Secco diluted (reference) _____ 86

Figure 65: A single etch pit found after etching with OPE B. _____ 86

Figure 66: Optical micrograph showing a high density of etch figures found in the epitaxial-layer after etching with the OPE B for 12 h.Removal: ~ 400 nm, magnification: 1000 times. This wafer fragment was partly covered with wax to determine the removal hence the appearance of artefacts on the wafer surface. _____ 87

List of illustrations

Figure 67: CZ-Bulk (NPC Fresh, Supl. 2) after etching with the Secco solution. Etching time: 16 minutes, removal: 12μm. A characteristic pattern on the wafer surface, the so called "flow-pattern", can be clearly seen. _____ 89

Figure 68: Square-shaped etch pits in CZ-Bulk (NPC Fresh, Supl. 2) after etching with OPE C. Etching time: 24h, removal: ~ 2μm. _____ 89

Figure 69: Structure of SOI. _____ 90

Figure 70: Effect of the dip in HF. _____ 92

Figure 71: Diameter (r) of the etched figures as a function of the hydrofluoric acid content. The diameter increases with increasing HF content. _____ 93

Figure 72: The diameter of the etched figures as a function of etching time shown for the OPE B. _____ 94

Figure 73: Standard Smart-Cut^{TM} SOI material after etching with the Secco diluted 1 reference. Etching time: ~ 3 min., initial thickness: 145 nm, thickness after etching: ~ 30 nm, subsequent dip in HF: 60s, magnification: 500 times. Defects are delineated as small circular etched figures. _____ 95

Figure 74: Standard Smart-Cut^{TM} SOI material after etching with the OPE F. Etching time: ~ 60 minutes, initial thickness: 145 nm, thickness after etching: ~ 30 nm, subsequent dip in HF: 60s, magnification: 500 times. The etched figures look similar to those delineated by the Secco diluted reference. _____ 95

Figure 75: Standard Smart-Cut^{TM} SOI material after etching with the OPE C. _____ 96

Figure 76: Comparison of the etched figure densities. _____ 97

Figure 77: SOI material with a high OSF density. This fragment was etched with the Secco diluted 1. Etching time: 90 s, subsequent dip in HF: 45 s, magnification: 1000x. The OSFs are revealed as oval shaped double pits. All etched figures have roughly the same length and shape. _____ 98

Figure 78: Oxidation-induced stacking faults in a SOI fragment etched with OPE F. _____ 98

Figure 79: The OPE C is also able to reveal oxidation induced stacking-faults. _____ 99

Figure 80: Different etched figures found in a SOI fragment etched with OPE D. _____ 100

Figure 81: SEM image of an etch pit found after treatment with OPE D. _____ 100

Figure 82: OSF densities found after etching with the Secco diluted reference and the OPE C and F. _____ 101

Figure 83: Defect densities found after etching with the Secco diluted reference and the various OPE mixtures. _____ 102

Figure 84: The samples were taken from the centre and the edge of the wafer. _____ 103

List of illustrations

Figure 85: Thin SOI after etching with the Secco diluted 1 solution. _____ 104
Figure 86: Thin SOI after etching with the OPE D. _____ 104
Figure 87: Crystalline defects in thin SOI material delineated with OPE A. _____ 105
Figure 88: The defect densities obtained after etching with different formulations. _____ 106
Figure 89: Swirl defects found after etching with the Secco diluted 1. Etching time: 10s, removal ~ 10 nm, magnification: 1000 times. _____ 106
Figure 90: Swirl defects revealed with the OPE A. _____ 107
Figure 91: Swirl-defect densities obtained after etching with the OPE A and OPE D. _____ 108
Figure 92: SIMOX SOI material after etching with the Secco diluted 1 _____ 109
Figure 93 : Crystal defects in SIMOX SOI found after etching with OPE F. _____ 109
Figure 94: Crystal defects in SIMOX SOI found after etching with OPE C. _____ 110
Figure 95 : Defect densities in SIMOX material found after etching with the OPE mixtures C, D and F. _____ 111
Figure 96: Depth distribution of crystal defects in the SOI layer. _____ 112
Figure 97 : Small etched figures found after etching with the OPE D. _____ 112
Figure 98: Standard Smart-Cut^{TM} SOI, etched with OPE D for 25 min., initial thickness: ~ 88 nm, thickness after etching: ~ 42 nm, magnification: 1000 times. _____ 113
Figure 99: The defect density increases with increasing removal. Even small defects are revealed by the preferential etch. Material used: Standard Smart-Cut^{TM} SOI, etching solution used: OPE D, etching time: 35 min., initial thickness: ~ 88 nm, thickness after etching: ~ 30 nm, magnification: 500 times. _____ 113
Figure 100: Defect density as a function of removal. _____ 114
Figure 101: Standard Smart-Cut^{TM} SOI material after etching with the OPE A. _____ 115
Figure 102: AFM image of an etch hillock (scan area 10 μm x 10 μm) found in standard Smart-Cut^{TM} SOI-material after etching with the OPE A. _____ 116
Figure 103: Standard Smart-Cut^{TM} SOI material after etching with the OPE B. _____ 116
Figure 104: AFM image of an etch hillock (scan area 10 μm x 10μm) found in standard Smart-Cut^{TM} SOI material after etching with the OPE B. Initial thickness of the SOI layer: ~62 nm, removal: ~ 17 nm. _____ 117
Figure 105: Standard Smart-Cut^{TM} SOI material after etching with the OPE D. _____ 117
Figure 106: AFM 3D image and line scan of an etch pit (scan area 10 μm x 10 μm) found in standard Smart-Cut^{TM} SOI material after etching with the OPE D. _____ 118
Figure 107: Etched figures in thick SOI material found after etching with the OPE D. _____ 119

List of illustrations

Figure 108: AFM 3D image and line scan of an etch hillock (scan area 20 μm x 20 μm) found in thick Smart-Cut™ SOI material (high Ω) after etching with the OPE D. _____ 119

Figure 109: SEM images of etch hillocks found in thick SOI after etching with the OPE D. 121

Figure 110: SEM image of a vertical cross-section of a selected etch hillock made with the FIB-technique. A cavity was found below the hillock. _____ 121

Figure 111 a-c: 200 kV X-TEM bright field image of an etch hillocks in SOI material. _____ 122

Figure 112: Preferential attack at the crystal defect and channel formation. in the SOI film. 123

Figure 113: Dissolution of the underlying buried oxide and hillock formation. _____ 123

Figure 114: Structure of sSOI _____ 124

Figure 115 Strained silicon on a virtual substrate _____ 125

Figure 116: Relaxing of the strained silicon layer _____ 125

Figure 117 Standard sSOI after etching with the OPE A. _____ 126

Figure 118: SEM image of standard sSOI after etching with the OPE A. _____ 127

Figure 119: Thin sSOI fragment after etching with the Secco dil. 2 reference. _____ 128

Figure 120: A sSOI sample from the same wafer as above after etching with the OPE F. _ 128

Figure 121: Thin sSOI material after etching with the OPE B. _____ 129

Figure 122: Experimentally determined TD densities. _____ 130

Figure 123: Indentations made by pressing a certain weight against the wafer surface. The indentations cause mechanical damage and strain inside the crystal lattice. A following annealing step leads to the formation of dislocations, primarily in the form of half loops. _____ 132

Figure 124: An indentation in the centre of a wafer fragment. Maximum force: 1N, magnification: 1000 times. This fragment was not etched. _____ 133

Figure 125: SEM image of oval-shaped etch pits caused by dislocations. _____ 134

Figure 126: The Secco diluted 1 solution is also able to reveal dislocations. _____ 135

Figure 127: MEMC solution: Pairs of oval-shaped pits found at dislocation half loops. _____ 135

Figure 128: OPE C revealed oval-shaped pits (SEM). Etching time: 12 h, removal at defect-free site: ~ 950 nm, magnification: 10 000 times. _____ 136

Figure 129: The OPE D has also the capability to reveal crystal defects. Dislocations are also delineated as oval shaped pits. Etching time: 8 h, removal at defect-free site: ~ 850 nm, magnification: 10 000 times. _____ 136

Figure 130: The OPE F produced pairs of polygonal etch pits at dislocation half loops. Etching time: 5 h, removal of perfect material: ~ 520 nm, magnification: ~ 25 000 times. _____ 137

List of illustrations

Figure 131: Experimental determination of the depth of a pit by AFM. The maximum depth that can be determined depends on the tip angle α. 137

Figure 132: The maximum determinable depth of an etch pit by AFM is dependent on the geometry and depth of the pit. 138

Figure 133: AFM 3D image (scan area 10μm x 10μm) and AFM line scan of two single pits found after etching with the Secco solution. 139

Figure 134: AFM 3D image (scan area 10μm x 10μm) 139

Figure 135: Dependence of the selectivity on etching time determined for the Secco solution. The selectivity apparently decreases with increasing removal. 141

Figure 136: Selectivities of the different etching solutions used. 144

Figure 137: Experimental arrangement used for the determination of the standard potential. 146

Figure 138: Selectivity as a function of activation energy for the etching process of silicon. 148

Figure 139: Selectivity as a function of removal rate. Shown for the Original Secco solution, the two different Secco diluted formulations and the MEMC and Jeita solutions. 148

Figure 140: Selectivity dependent on removal rate experimentally determined for the OPE C, D and F. 149

Figure 141: Selectivity as a function of standard potential. 149

Figure 142: Activation energy for the etching process of silicon as a function of the standard potential of the etching solution. 150

Figure 143: A comparison between the defect densities of the Secco diluted 1 reference and the OPE F 152

1 Introduction

Monocrystalline silicon rods (ingots) of low defect density and a large diameter are used as the base material for microelectronic devices. For further processing the silicon ingot is cut into slices with a diameter ranging from 150 mm to 300 mm and with a thickness of ~ 750 µm. These perfectly circular and flat disks are called **wafers**. After going through several processing steps including chemical etching, polishing and cleaning it is possible to build microelectronic devices like integrated circuits (IC´s) onto the wafer surface. Integrated circuits are used extensively in all walks of live. They appear in mobile phones, cars or even in simple kitchen appliances like coffee machines or toasters, for example.

The integrated circuit was invented independently in the late fifties of the last century by **Jack Kilby** of Texas Instuments, Inc. and **Robert Noyce** of Fairchild Semiconductor Corporation. An integrated circuit is an assembly of electronic components, fabricated as a single unit, in which miniaturized active devices like transistors and diodes and passive devices like capacitors or resistors and their interconnections are built up on a thin substrate of semiconductor material which is typically silicon. The resulting circuit is thus a small monolithic "chip", which may be as small as a few square centimetres or only a few square millimeters. The individual circuit components are generally microscopic in size [1]. Integrated circuits are used to create microprocessors or memory cells for example. Transistors are the most important components of integrated circuits. Gordon Moore, the co-founder of Intel, predicted in 1965 that the number of transistors on a chip would double every two years. This rule is also known as **Moore´s law** and has been valid over the last forty years. The number of transistors in an integrated circuit could be increased by decreasing their size. In the early sixties of the last century an integrated circuit contained only a few transistors. Fifteen years later an integrated circuit already consisted of a few thousand transistors (large-scale integration). It then became possible to integrate a microprocessor on one chip.

With the beginning of the very-large-scale integration (VLSI) era in the early eighties of the last century it was possible to produce integrated circuits consisting of hundreds of thousands of transistors. The microprocessors used today contain millions of individual transistors [2]. The integration of millions of transistors on one chip was possible by reducing the size of IC devices to nanometer scale. As transistors are the most important units in semiconductor devices their mode of operation will be discussed briefly.

There are two basic types namely the bipolar or carrier injection transistor and the unipolar or field-effect transistor.

The metal oxide semiconductor field-effect transistor (MOSFET) is normally used in microprocessors or memory circuits. It contains three different electrodes: the source (S), the gate (G) and the drain (D). The gate electrode consists of heavily doped polysilicon or metals like aluminum for example.

The source and drain are islands of highly n-doped (n^+) silicon surrounded by the p-type silicon substrate (NMOS) or vice versa (PMOS). In the NMOS, electrons are the majority carriers in the n^+ source and drain regions. They are separated by a channel of p-type silicon where holes are the dominating carriers. The gate electrode is above this small channel. The gate electrode and the underlying substrate are separated by an insulator which is normally silicon dioxide (figure 1).

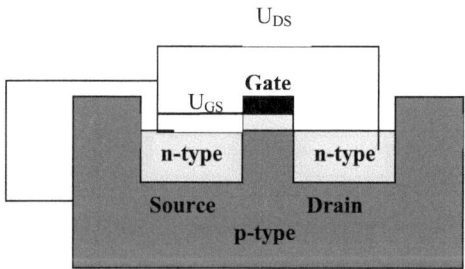

Figure 1: Construction of an n-doped metal oxide semiconductor field-effect transistor (NMOS).

A potential difference U_{DS} exists between the source and the drain.

When no voltage is applied to the gate ($U_{GS} = 0$) no current flows between source and drain.

When a positive bias voltage ($U_{GS} > 0$) is applied to the gate a current flows if U_{GS} exceeds a characteristic limit called threshold voltage (U_{th}).

Minority carriers (electrons) move from the p-doped silicon substrate to the area below the gate electrode. Source and drain are now connected by a conducting n-channel through which a current can flow (figure 2). The conductance of this channel and therefore the current flow I_D can be modulated by varying the gate voltage ($I_D \approx U_{GS}$). This is a characteristic of MOSFET [3].

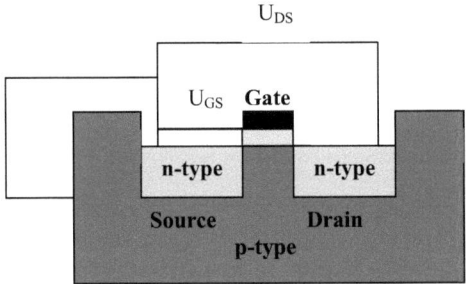

Figure 2: Operating NMOS (UGS > Uth). A conducting channel is formed between source and drain. A current flows from source to drain.

It is possible to store information by using modified MOS-transistors which have two gates instead of one. The additional gate, called the floating gate, is electrically isolated from the control gate (figure 3).

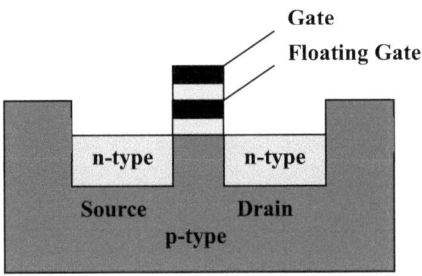

Figure 3: Structure of a floating gate metal oxide semiconductor field-effect transistor (FGMOSFET). The floating gate is separated from the control gate by a thin silicon dioxide layer.

When a positive bias voltage ($U_{GS} \approx 5$ V) is applied to the control gate a current flow between source and drain is obtained. The floating gate is uncharged. When a high positive bias voltage ($U_{GS} \approx 10-15$ V) is applied to the control gate electrons can jump through the insulating layer onto the floating gate. The floating gate is charged. No current can flow between source and drain. This process is called hot-electron injection and may be explained by the tunnel effect.

The initial state of the floating gate can be restored by applying a high negative bias voltage($U_{GS} \approx$ -10– 15 V) to the control gate. The charged state is the "1" state (= yes). The uncharged state is synonymous with the "0" state (= no).

Floating gate transistors are used in EPROM or EEPROM memory cells.

Silicon produced for microelectronic applications is the purest and the most perfect crystalline material manufactured today. Nevertheless, this nearly perfect material contains different defects which are introduced into the silicon crystal during the growth process and further processing. Crystalline defects in substrates for microelectronic devices are normally undesired. They can have a negative impact on the functionality and reliability of integrated circuits (IC´s). To characterize the quality of the substrates it is necessary to reveal the different crystal defects. The defect density per square cm or per cubic cm is a very important measure of the quality of a substrate wafer. The different crystal defects are normally delineated by using a so-called preferential etching solution. Such solutions are able to reveal crystal defects. After etching crystal defects are normally delineated as pits and sometimes also as hillocks.

Various etching solutions for defect delineation have been employed. They generally contain a strong oxidizing agent that converts the silicon formally into SiO_2, hydrofluoric acid which dissolves the oxidation product and a dilution component like water or acetic acid. Most of the recipes used today contain a chromium (VI) species like CrO_3 or an alkali dichromate. Such chromium based recipes are highly suited to revealing different crystal defects in the substrate wafers and diluted versions can be used for defect delineation in thin silicon films. However, chromates, and dichromates in particular, are highly toxic and carcinogenic due to their ability to interact with cells and the DNA [4].

Chromium (VI) compounds are also harmful to the environment. They are very toxic to aquatic organisms and may cause long-term adverse effects in the aquatic environment [5]. It is obvious that the use of Cr (VI) has to be restricted.

A new class of preferential etches for defect delineation should ideally have the following characteristics:

- the formulation should be free of chromium (VI) compound to reduce the risks to health and the environment
- low etching rates so that even thin silicon films can be etched with a sufficient control of the etching time [4]

- etch homogeneity
- etch sensitivity, i.e. the ability to reveal and to differentiate between the different crystal defects
- stability of the etching solution, so that the same can be stored for a certain period of time and can be handled without highly elaborated safety measures [4]

In chapter 2 the production of silicon wafers based on quartz (SiO_2) as the starting material will be discussed in detail. Novel silicon substrates like silicon on insulator (SOI) and their properties will also be discussed.

The different crystal defects which arise during the various processing stages are described in chapter 3.

The existing chromium-based and chromium-free formulations which can be used for the delineation of crystal defects in silicon substrates or thin films will be discussed in chapter 4. General aspects of the etching mechanism of silicon are also given in chapter 4.

A new class of environmentally friendly chromium free etching solutions – the Organic Peracid Etches – are introduced in chapter 5. The Organic Peracid Mixtures (OPE) provide very low removal rates. Therefore it is possible to use them for the delineation of defects in thin and very thin silicon films. They are also able to reveal different types of crystal defects.

In chapter 6 selected existing chromium-based and chromium-free formulas and the new Organic Peracid Mixtures are compared with respect to their chemical and physical properties.

A short summary of the experimental results and an outlook are finally given in chapter 7 and 8 respectively.

2 Crystal growth process and wafer manufacturing

2.1 Introduction

Monocrystalline silicon wafers are the base material for microelectronic devices and circuits. Silicon is the second most abundant element on earth. More than 90% of the earth´s crust consists of silica or silicate [3].
Semiconductor grade silicon wafers are required to have a purity of 99.999999999 % regarding contaminations by dopants and metals. The oxygen content can range from 10 to 20 ppm depending on the crystal growth process (chapter 2.3)

The starting material is quartz. Several chemical, mechanical and thermal processes are necessary to first convert quartz (SiO_2) into polycrystalline silicon with a purity of 99.999999999 %. Two techniques, the Float-Zone (FZ) method and the Czochralski (CZ) method, are used to convert polysilicon into a monocrystalline silicon ingot. Out of the silicon ingot by several steps which involve slicing, polishing, etching and cleaning, the final wafers are manufactured. Table 1 shows the different process steps which are necessary to produce silicon wafers.

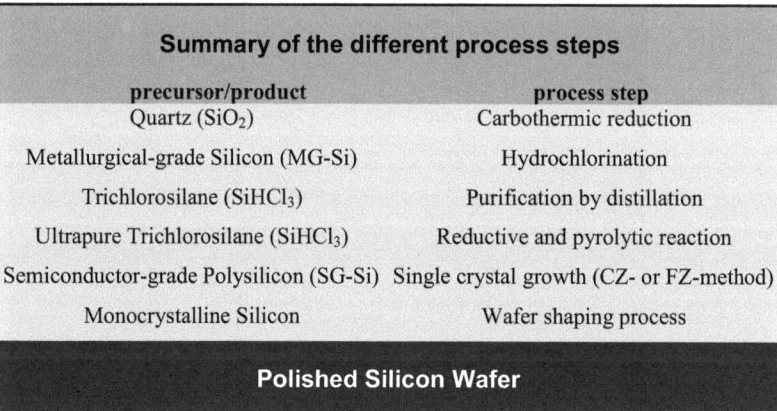

Summary of the different process steps	
precursor/product	process step
Quartz (SiO_2)	Carbothermic reduction
Metallurgical-grade Silicon (MG-Si)	Hydrochlorination
Trichlorosilane ($SiHCl_3$)	Purification by distillation
Ultrapure Trichlorosilane ($SiHCl_3$)	Reductive and pyrolytic reaction
Semiconductor-grade Polysilicon (SG-Si)	Single crystal growth (CZ- or FZ-method)
Monocrystalline Silicon	Wafer shaping process
Polished Silicon Wafer	

Table 1: Summary of the stages in wafer manufacturing.

2.2 Preparation of high-purity silicon

The starting material for high-purity silicon single crystals is silica (SiO_2).
This is firstly reduced by carbon, normally in the form of coal and coke, to elemental Si in an arc furnace:

(1) $SiO_2 + 2C \rightleftharpoons Si + 2CO$

Many different reactions occur at different temperatures in the furnace:

(a) $SiO_2 + C \rightleftharpoons SiO + CO$ (1700 - 1900°C)
(b) $SiO + 2C \rightleftharpoons SiC + CO$ (1500 - 1700°C)
(c) $SiO_2 + SiC \rightleftharpoons SiO + Si + CO$ (>2000°C)
(d) $SiO + CO \rightleftharpoons SiO_2 + C$ (<1500°C)

The silicon which is obtained in this way is referred to as metallurgical grade silicon (MG-Si). It contains about 1 -2 mass percent of impurities which are mainly metals like iron and aluminium. Metallurgical grade silicon also contains non-metals like boron or phosphorus in the parts per million (ppm) range [6].
To produce 1 metric ton of MG-Si 2.9 – 3.1 t silica, 1.2 – 1.4 t coke, 1.7 – 2.5 t coal or wood chips and 12.5 – 14 kWh are necessary [6].
The next step is to purify the MG-Si to the level of semiconductor grade silicon (SG-Si). Metallurgical grade silicon powder is made to react in a fluidized-bed reactor with anhydrous HCl and forms several chlorosilanes (figure 4).
The main as well as the desired product is trichlorosilane ($SiHCl_3$). But monosilane (SiH_4), silicon tetrachloride ($SiCl_4$) and dichlorosilane (SiH_2Cl_2) are also formed [3].

(2) $Si + 3HCl \rightleftharpoons SiHCl_3 + H_2$ (400°C)

Metallic impurities like Fe and Al form chlorides like $FeCl_2$ and $AlCl_3$ which can be removed easily [6].
The trichlorosilane ($SiHCl_3$) which has a low boiling point (32°C) can be removed from the reaction mixture by fractional distillation (figure 4).

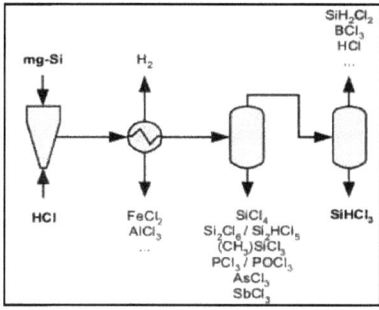

Figure 4: Formation of SiHCl$_3$ followed by fractional distillation.
Reference: *Chemie Unserer Zeit (37), 2003*

In the deposition reactor the high purity trichlorosilane is then vaporized and mixed with high-purity hydrogen. The silicon is deposited at thin heated (1100°C) silicon rods within the deposition reactor (figure 5). This process is referred to as the Siemens Process. [3].

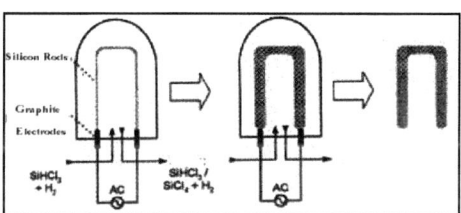

Figure 5: The SiHCl$_3$ is decomposed to polycrystalline silicon at the heated U–shaped silicon rods.
Reference: *Chemie Unserer Zeit (37), 2003*

Different reactions occur [6]:

(3) $4\ SiHCl_3 \rightleftharpoons Si + 3\ SiCl_4 + 2\ H_2$

(4) $SiCl_4 + H_2 \rightleftharpoons SiHCl_3 + HCl$

The ultra-pure semiconductor grade silicon thus deposited contains impurities that range from parts per trillion (ppt) to parts per billion (ppb).

The next step is the growth of single crystals. Two techniques dominate the production of silicon single crystals. One is the Float-Zone (FZ) method and the other is the Czochralski (CZ) pulling method.

About 95% of single crystal silicon is produced by the CZ method.

2.3 Crystal growth methods

2.3.1 The Czochralski Method

The semiconductor grade polycrystalline silicon is heated under a protective atmosphere in a quartz crucible to temperatures above the melting point of silicon which is 1420°C. The quartz crucible is surrounded by a graphite crucible which acts as a resistance heater (figure 6). The silicon is kept at a high temperature for a while to ensure complete melting and to avoid the formation of tiny bubbles. An inert gas like argon is passed through the melt to carry away reaction products like SiO from the quartz crucible and CO from the graphite crucible. Oxygen and carbon are the main impurities in the Czochralski silicon monocrystals (chapter 3).

A thin seed crystal is then dipped into the melt until it begins to melt itself [3]. The crystal growth starts at the seed crystal which is then drawn from the melt. The orientation of the growing crystal is determined by the orientation of the seed crystal which rotates during the process (figure 6). The diameter of the growing crystal is dependent on the pulling rate and on the temperature of the melt. A decrease in the pulling rate or the temperature of the melt increases the diameter of the growing crystal.

At the beginning of the pulling process dislocations occur in the growing crystal due to the thermal shock. To avoid this the Dash technique is used. At first the crystal is grown with a relatively high pulling rate which leads to the formation of a narrow neck. This process is called "necking". The pulling rate is higher than the diffusion controlled growth rate of the dislocations which end at the surface of the thin neck [7].

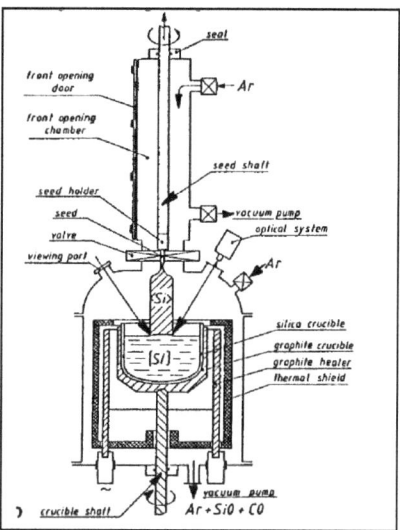

Figure 6: A typical Czochralski silicon crystal growing system.
Reference: *http://www.fullman.com/semiconductors*.

The formation of a neck is absolutely necessary to produce dislocation free crystals. This technique is used for both the Float-Zone (FZ) and the Czochralski (CZ)-method [3].

After the necking process the temperature of the melt is decreased. The diameter of the growing crystal increases forming a cone and then a shoulder and finally the cylindrical body with a constant diameter (figure 7).

Towards the end of the growth process the crystal diameter is allowed to reduce gradually and end in a cone to minimize thermal shock which can cause dislocations at the end of the crystal [3].

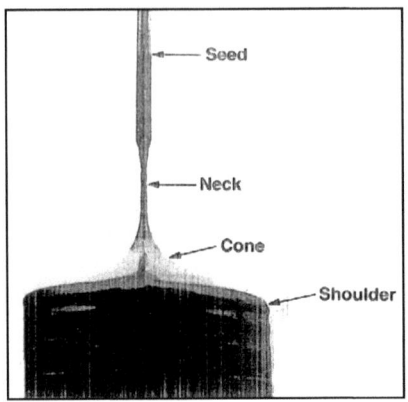

Figure 7: A Czochralski grown silicon crystal.
Reference: *A. Shimura Semiconductor Silicon Crystal Technology.*

The shoulder and ends are not used for device fabrication as their crystal properties are not homogeneous owing to the abrupt change in growth conditions. Dopants may be added to the silicon melt to obtain the desired resistivity. However, as the physicochemical properties of pure dopants are often quite different from those of the silicon melt and, except for heavily doped material, the amount of dopant used is too low to ensure an even distribution in the silicon bulk, they are added to the melt as highly doped silicon particles.

Czochralski grown silicon crystals are more uniform than those produced by the Float Zone method regarding dopant distribution [3].

2.3.2 The Float-Zone Method

A polysilicon rod is made to melt at one end with a movable induction coil. The molten tip is brought into contact and fused with a single crystal seed which has the desired orientation. The induction coil is then moved along the rod melting the zone it surrounds (figure 8). The molten zone solidifies as the induction coil moves away and grows as an extension of the seed crystal. The silicon rod rotates in a protective atmosphere while the melting zone moves from one end of the rod to the other (figure 8).

Just as in the CZ process after seeding a thin neck of about 2 or 3 mm in diameter and with a length of 10-20 mm is formed to avoid dislocations which occur in the newly grown single crystal during the seeding process due to the thermal shock.

In the case of the CZ process the silicon melt is in contact with a quartz crucible which results in a relatively high oxygen concentration (10^{18} atoms/cm^3) [3].

FZ crystals have a much higher level of purity than CZ grown crystals as the molten silicon only comes into contact with the non-reactive ambient gas in the reaction chamber.

The oxygen concentration for example is in the range of 10^{16} atoms/cm^3 (chapter 3). The impurities present are less soluble in the crystal than in the melt and are therefore drawn with the melting zone to the end of the rod resulting in a silicon crystal with a very low concentration of foreign atoms. However for technical reasons the diameter of the single crystal ingot is much smaller than that of the CZ produced ingot making the FZ process less cost-effective. FZ silicon is a high purity alternative to Czochralski grown silicon whereas the concentration of carbon and oxygen impurities are extremely low [8].

The purity of the FZ material leads to a resistivity of 200 Ω cm ore even more. Therefore FZ silicon can be used for power devices that must support reverse voltages up to 1000 V [3]. FZ material is also used for high efficiency solar cells and RF wireless systems.

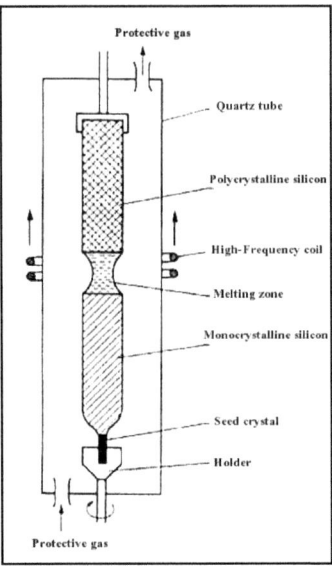

Figure 8: The Float-Zone process.
Reference: *H. Kück izfm (Universität Stuttgart)*

Gaseous dopants may be added to the silicon melt. A dopant gas such as phosphine (PH_3) for n-type silicon or diborane (B_2H_6) for p-type silicon is blown into the reaction chamber during the growth process [3].

2.4 Wafer manufacturing

2.4.1 Slicing and polishing

The seed and tail ends of the single crystal ingot are then cut off.

The silicon ingot is then mechanically ground to a cylindrical shape of a precise diameter. In one or more parts the edge is flattened along the length of the cylinder to identify the crystallographic orientation.

The mechanical damages induced by grinding and marking are removed by chemical etching, generally with mixtures containing HNO_3 and HF [3].

Wafers of thicknesses varying from 200 to 750 μm are cut from the silicon ingot using either a diamond saw (fig. 9) or a multiple wire saw (fig. 10).

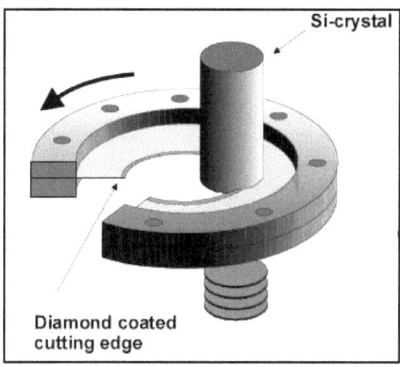

Figure 9: Slicing of wafers with an annular or internal diameter saw.
Reference: *IMTEC*

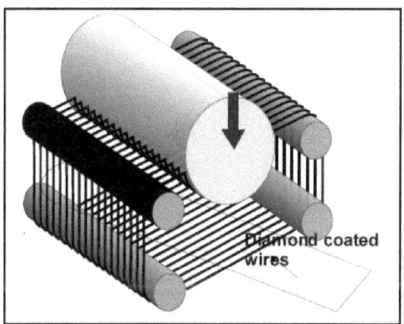

Figure 10: Slicing with a multiple wire saw.
Reference: *IMTEC*

In a multiple wire saw the cylindrical ingot is surrounded by several parallel thin metal wires which by rapid rotation cut slices off the wafer (figure 10).

After slicing, the edges of the silicon wafers are rounded. Damages to the wafer surfaces caused by slicing can be removed by lapping with a slurry which normally contains aluminium oxide or silicon carbide and glycerine. After lapping a uniformly flat and planar surface is obtained

The damage caused by slicing and lapping reaches down to a depth of 10µm

and is removed by chemically etching off the surface to a depth of 10 – 30 µm in an acid or alkaline bath.

An alkaline etching solution contains 10 – 50 % KOH or NaOH that dissolves the silicon according to the equation:

(5) $Si + 2\,H_2O + OH^- \rightleftharpoons HSiO_3^- + 2\,H_2$

An acid based etching solution always contains an oxidizing agent like nitric acid, a diluent which may be water or acetic acid, and hydrofluoric acid.

Silicon is first oxidized to SiO_2 which is then dissolved by HF according to the following equations:

(6) $Si + 4\,HNO_3 \rightleftharpoons SiO_2 + 2\,H_2O + 4\,NO$

(7) $SiO_2 + 6\,HF \rightleftharpoons H_2SiF_6 + 2\,H_2O$

The wafers which have been treated with one of the abovementioned etching solutions are then rinsed with ultra pure water and polished with a soft cloth to give a plane and smooth surface.

2.4.2 Wafer cleaning

During the various stages of slicing, grinding, and polishing, contaminants of a metallic, organic and particulate nature accumulate on the wafer. The various contaminants can be removed by using different chemical cleaning procedures. At first an alkaline cleaning step called "Standard Clean 1 (SC1)" is used to remove particles and organic compounds from surface (table 11).The wafers are immersed for 10- 20 minutes into a tank containing the SC1 solution maintained at 70-80°C. After rinsing with DI H_2O an acid based cleaning step called "Standard Clean 2 (SC 2)" followed by HF treatment is used. The SC 2 solution is normally also maintained at 70-80°C [9]. Finally the wafers are rinsed again with DI water. For the Standard Clean 1 a solution containing DI H_2O, H_2O_2 and NH_3 is normally used. The hydrogen peroxide is electronic grade without any stabilizers to prevent contamination of the cleaning solution and the wafer surface [9]. Instead of NH_3 other bases like tetramethyl - ammonium hydroxide (TMAH) can also be used. The 1 -2 nm thick native oxide on the top of the surface is dissolved taking with it organic contaminants and particulate matter which would have been adsorbed to the surface. A new oxide which reaches into the silicon crystal is formed. Transition metals like Cu, Au, Ag, Zn, Cd, Ni, Co or Cr are also partially dissolved forming amine complexes [9]. The Standard Clean 2 solution normally contains DI H_2O,

H_2O_2 and HCl and is used to remove metals from the wafer surface. It is able to remove alkali residues and residual trace metals like Au and Ag. Metal hydroxides like $Al(OH)_3$, $Fe(OH)_3$, $Mg(OH)_2$, $Ca(OH)_2$ or $Zn(OH)_2$ are dissolved by forming chlorides [9].

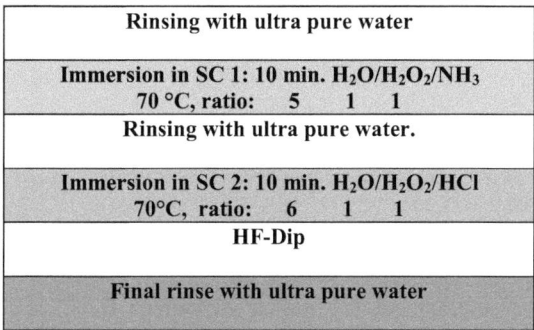

Table 2: Stages in the cleaning of silicon wafers.

After the SC1 and SC2 step a hydrophilic surface is obtained (figure 11). An additional dip in HF can be used finally to generate a H-passivated, hydrophobic silicon surface (figure 11).

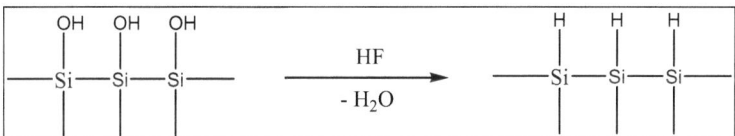

Figure 11: Formation of a hydrophobic surface after HF treatment and elimination of water.

2.5 Epitaxy

The cleaned and polished silicon wafers have a roughness in the atomic range. This means that they are much smoother than the best mirror. Such wafers are suitable for most applications. But it is possible to refine the silicon wafers once again by growing a thin crystalline layer on the top of a substrate (figure 12). This process is called Epitaxy [6].
If the thin layer is grown on the same substrate like silicon on silicon the process is called homoepitaxy. If the thin layer and the substrate are of different materials like silicon and germanium for example the process is called heteroepitaxy [3].

The epitaxial layer with a thickness of 10 nm – 100 μm has the same crystal properties as the underlying substrate.

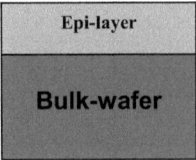

Figure 12: Epitaxial layer grown on a substrate wafer by chemical vapour deposition (CVD).

The epitaxial layer is grown by chemical vapour deposition (CVD). Gaseous compounds like $SiCl_4$, $SiHCl_3$, SiH_2Cl_2 or SiH_4 and possibly hydrogen are mixed with a carrier gas in a reactor (table 2). The substrate wafer lies on a susceptor inside the reaction chamber (figure 13). The SiC coated susceptor and the reaction chamber are heated by radio–frequency induction (figure 13).

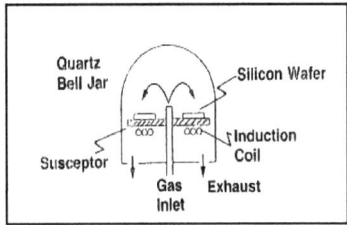

Figure 13: Vertical epitaxy reactor.
Reference: A. Shimura Semiconductor Silicon Crystal Technology.

The temperature inside the reaction chamber ranges from 950°C – 1250°C (table 2). At these temperatures the gaseous silanes decompose and silicon is formed which is deposited on the surface of the wafer.

Compound	Growth rate (μm/minute)	Temperature range (°C)
SiH_4	0.2 – 0.3	950 - 1050
SiH_2Cl_2	0.4 - 3	1050 - 1150
$SiHCl_3$	0.4 - 2	1100 - 1200
$SiCl_4$	0.4 – 1.5	1150 - 1250

Table 3: Different silanes which can be used for the epitaxy process.

The formation of the silicon is not as easy as it seems to be:

(8) $SiCl_4 + 2 H_2 \rightleftharpoons Si + 4 HCl$

(9) $SiHCl_3 + H_2 \rightleftharpoons Si + 3 HCl$

(10) $SiH_2Cl_2 \rightleftharpoons Si + 2 HCl$

The course of reaction is relatively complex. Several intermediates are formed:

(11) $SiCl_4 + H_2 \rightleftharpoons SiHCl_3 + HCl$

(12) $SiHCl_3 + H_2 \rightleftharpoons SiH_2Cl_2 + HCl$

(13) $SiH_2Cl_2 \rightleftharpoons SiCl_2 + HCl$

(14) $SiCl_2 + H_2 \rightleftharpoons Si + 2 HCl$

For the formation of silicon from SiH_4 only two steps are known [3].

(15) $SiH_4 \rightleftharpoons SiH_2 + H_2$

(16) $SiH_2 \rightleftharpoons Si + H_2$

Dopants can be added to the CVD system during the entire growth process. They are added as hydrides like B_2H_6 or PH_3 for example [3].

The growth rate of the epitaxial layer is dependent on deposition temperature, pressure (concentration) and the type of gas used for the process [3].

The oxygen content has to be below 1 ppm [3].

The new epitaxial layer is nearly defect free. Only imperfections at the substrate surface like contaminants or imperfections in the crystal lattice lead to the formation of stacking faults in the epitaxial layer [6].

2.6 New materials

2.6.1 Introduction

The increasing demand for portable systems capable of multimedia functions has driven the IC industry to develop low power, low voltage and high performance circuits.
Novel substrate materials like Silicon On Insulator (SOI), have already been introduced while Strained Silicon On Insulator (sSOI) or Germanium On Insulator (GeOI) are being developed to overcome the limitations of conventional silicon substrates or epitaxial wafers regarding further progress in the performance of microelectronic devices [10].

2.6.2 Silicon on Insulator (SOI)

SOI consists of a thin layer of silicon on top of an insulator, normally silicon dioxide (figure 14). Electronic devices will be built in this thin SOI layer. The basic idea is to reduce the capacitance of the transistor.

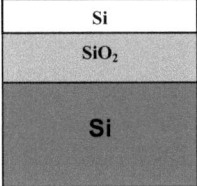

Figure 14: Structure of SOI wafer.

Most silicon wafers have a thickness of around 750 µm of which only a few micrometers at the surface are used for the fabrication of metal–oxide semiconductor (MOS) transistors. Interactions between device and substrate lead to undesirable effects, one of which is the parasitic capacitance between source, drain and substrate which lowers performance and increases energy consumption.
To avoid such effects the microelectronic devices are built in the thin silicon top layer which is separated by the insulator SiO_2 from the substrate (figure 14).

Today there are two different processes with which Silicon On Insulator (SOI) wafers are produced: SIMOX (Separation by Implanted Oxygen) and Smart Cut^{TM}.

The SIMOX technique consists of two steps, the formation of a buried oxide layer (BOX) by implantation of O^+ - ions and a high temperature annealing (figure 15). The thickness of the BOX layer is dependent on the oxygen dose and the temperature during the implantation process and the kinetic energy of the ions determine the depth of the buried oxide layer [11]. Normally a dose around 1.4×10^{18} O^+ ions/ cm^2 with an energy of 200 keV is used to generate a continuous oxygen layer. Oxygen distribution during the implantation process follows a Gaussian curve. A new process uses a low oxygen implantation dose which is about 4×10^{17} O^+ ions/cm^2. This new technique limits the buried oxide thickness to a range of 80 – 100 nm [12]. The new process increases the crystal quality by lowering defect density [13].

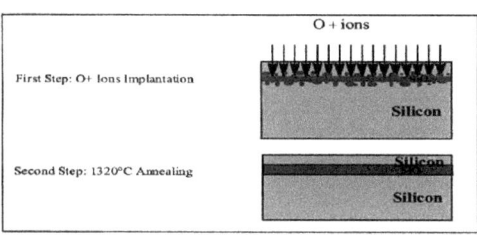

Figure 15: The SIMOX-process.
Reference: *A. J. Auberton Proceedings of the European Solid State Device Research Conference (IEEE), 1996*

After implantation high temperature annealing is necessary to form a uniform and stoichiometric buried oxide layer and to regenerate the crystal lattice of the silicon layer above the buried oxide [12]. The higher the temperature the better the quality of
the SOI layer [13].

The second method of producing SOI wafers is the Smart-Cut^{TM} process (Fig.16).

Two wafers are involved in the process. The surface of one wafer is thermally oxidized and functions as the buried oxide layer of the SOI structure later on. Hydrogen ions of the order of $2 \times 10^{16} - 10^{17}$ ions/cm^2 are then implanted below the oxide layer, the depth of penetration is depending on the kinetic energy of the hydrogen ions. The implantation process leads to imperfections in the crystal lattice arising from micro cavities and micro bubbles. Inside these cavities Si-H bonds are formed. The two wafers are cleaned and polished to produce a hydrophilic surface which is necessary for chemical bonds to be formed between the wafers. They are now bonded together as show in fig. 16.

The top wafer is then cut off by thermal activation at 500°C to 600°C. At this temperature Si-H bonds dissociate and hydrogen is set free. The micro cavities and bubbles merge to form a cleavage plane. The bonded wafer is now separated into a SOI wafer and a silicon wafer which can be reused for the next Smart Cut process. The SOI wafer is annealed at 1100°C to heal the crystal damages caused by the Smart Cut process and the surface is polished by oxidation and HF treatment.

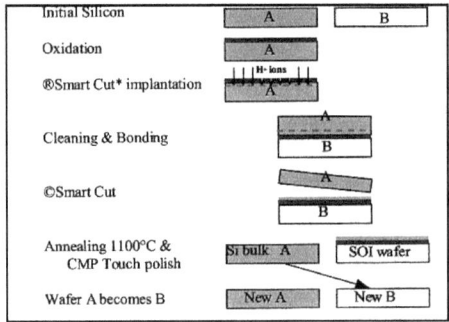

Figure 16: Smart-Cut process for SOI production.
Reference: *A. J. Auberton Proceedings of the European Solid State Device Research Conference (IEEE), 1996*

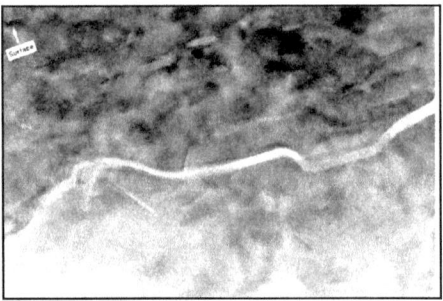

Figure 17: High resolution TEM image of the Smart-Cut region just before splitting.
Reference: *A. J. Auberton Proceedings of the European Solid State Device Research Conference (IEEE), 1996*

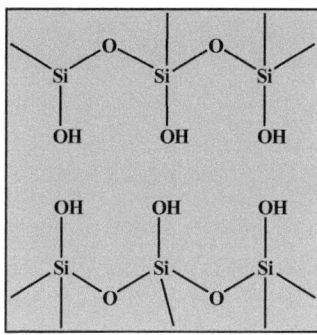

Figure 18: Hydrophilic interactions.

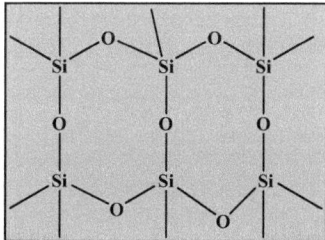

Figure 19: Formation of chemical bonds after annealing and elimination of water.

2.6.3 Strained Silicon on Insulator (sSOI)

Strained Silicon on Insulator (sSOI) consists of a thin strained silicon layer on top of an insulator like silicon dioxide (figure 20).

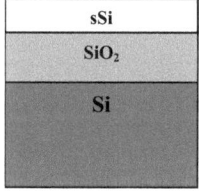

Figure 20: Structure of sSOI.

The strain in the crystal lattice alters the band structure as well as the electronic properties of the silicon. These changes result in an enhanced carrier mobility. The reduced capacitance caused by the SOI structure in combination with the increased carrier mobility leads to an appreciable performance enhancement [14].

To obtain sSOI wafers a technique which is similar to the SOI production is used.

Two wafers are bonded. One wafer contains the strained silicon (sSi) layer, the other one the SiO_2 layer.

A strained silicon layer is obtained by epitaxial silicon deposition on a relaxed SiGe buffer layer which is referred to as a virtual substrate (figure 21) and which is also obtained by epitaxial deposition.

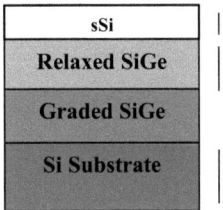

Figure 21: Strained silicon on a virtual substrate.

The SiGe buffer layer has a larger lattice constant (depending on the SiGe-alloy composition which may be between 0 and 4.2%). Due to its smaller lattice constant the silicon on top of the virtual substrate expands laterally and contracts vertically (figure 22).

The resulting biaxial stress enhances its carrier mobility [15].

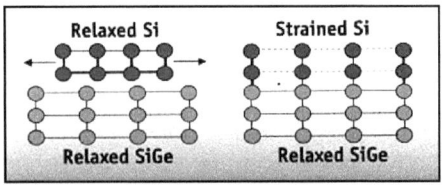

Figure 22: Crystal lattice of strained silicon.
Reference: *New Ideas for New Materials: Advanced Substrates and Devices for Nanoscale CMOS*

The strain inside the crystal lattice has been investigated by Micro Raman spectroscopy (figure 23). Unstrained silicon has a characteristic Raman peak at 520 cm^{-1}. The Raman peak of strained silicon appears at 514 cm^{-1} which indicates strain in the order of 0.8 % [14].

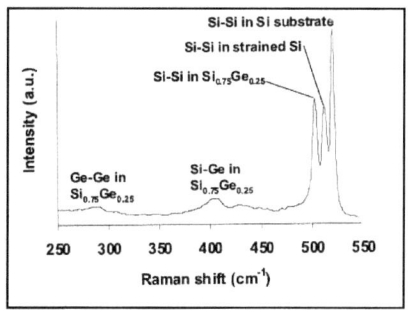

Figure 23: Raman spectrum of fabricated sSOI.
Reference: *S. H. Christiansen ATMI Epitaxial Services Mesa, AZ, USA*

For the removal of the virtual substrate after bonding hydrogen ions are implanted through the surface of the wafer with an energy of approximately 200 keV and a dose of 8 x 10^{16}/cm^2 [15]. After bonding the newly formed wafer is annealed (as a rule at 500 – 600°C for 2 hrs.). The annealing induces a splitting process which separates the bonded wafer pair parallel to the surface between the relaxed SiGe and the graded SiGe layers (figure 24) [14].
The 700 nm thick SiGe-layer remaining on the bonded wafer is removed by selective etching (figure 24) [14].

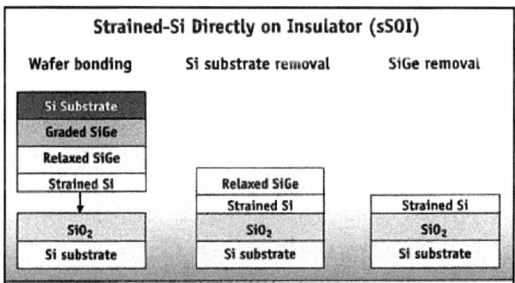

Figure 24: sSOI fabrication. After bonding the virtual substrate is removed by ion implantation followed by annealing and selectively etching.
Reference: *New Ideas for New Materials: Advanced Substrates and Devices for Nanoscale CMOS*

Various wet chemical etches can be used for this purpose. Mixtures consisting of HF, H_2O_2 and HAc, or NH_3, H_2O_2 and HF or HNO_3 and HF are normally used. These etching solutions have a high removal rate on SiGe and a very low removal rate on silicon or strained silicon [14] which means that the etching process can be stopped selectively at the sSi layer.

3 Crystal defects

3.1 Introduction

Silicon produced for microelectronic applications is the purest and the most perfect crystalline material manufactured today [16]. Nevertheless, this nearly perfect material contains different crystal defects which arise during the growth process described in chapter 2. These imperfections of the crystal lattice are often called "Bulk Micro Defects (BMDs)". BMDs include any kind of imperfections like vacancy agglomerates (D-defects), oxygen precipitates, dislocations, etc.

Crystal defects and impurities are normally not desired in semiconducting materials in the device active zone. They can create energy states in the bandgap, decrease the carrier lifetime or generate minority carriers and can also act as gettering sites at which metallic impurities tend to concentrate [17].

Crystal defects may be classified by their geometry (table 4 and figure 25).

Crystal defects

Geometry	Defect
Point	Intrinsic point defect: - Vacancy - Self-interstitial
Point	Extrinsic point defect: - Substitutional impurity atom - Interstitial impurity atom
Line	Dislocations: - Edge dislocation - Screw dislocation - Dislocation loop
Plane	- Stacking fault - Twin - Grain boundary
Volume	- Precipitate - Void (vacancy agglomerates)

Table 4: Different crystal defects

Figure 25 shows a simple two-dimensional model of a cubic crystal lattice. Different types of crystal defects are illustrated:

a) vacancy (one atom is removed from the crystal lattice)
b) self-interstitial (one similar atom is added to the crystal lattice)
c) extrinsic substitutional atom which, according to size, contracts or widens the crystal lattice
d) interstitial impurity
e) edge dislocation
f) dislocation loop formed by agglomeration of self interstitials
g) dislocation loop formed by agglomeration of vacancies
h) precipitate of impurity atoms
i) vacancy cluster (Void or D-defect).

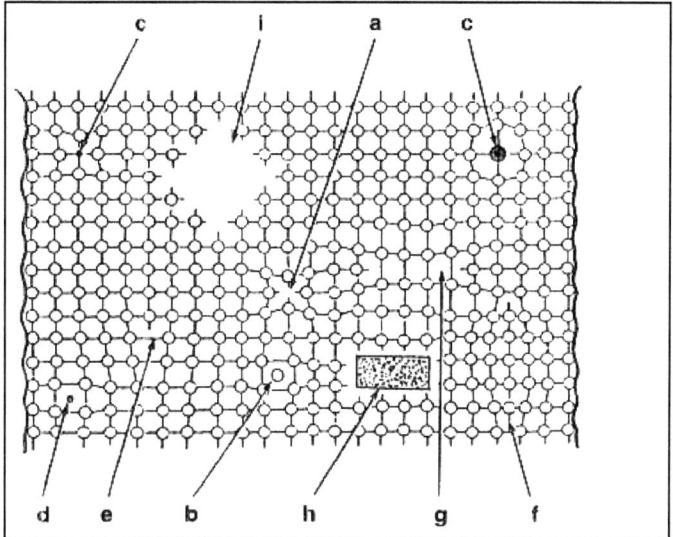

Figure 25: Simple two-dimensional model of a crystal lattice showing different types of crystal defects.
Reference: *A. Shimura Semiconductor Silicon Crystal Technology.*

3.2 Dislocations

Dislocations are imperfections or disturbances of an otherwise perfect crystal lattice [3]. Dislocations are caused by mechanical stress, elastic or plastic deformations, or by a large temperature gradient. In the case of an elastic deformation the crystal returns to its initial state when the causal force diminishes. If the force exceeds the yield strength the crystal is plastically deformed [3]. The mechanical stress inside the crystal is reduced by forming dislocations. There are two basic types of dislocations, namely edge and screw dislocations (figure 26 and 27). Any dislocation can be regarded as a superposition of an edge and a screw dislocation [18]. If an additional half- plane (of atoms or ions) is inserted into the crystal lattice a dislocation is formed. Where the extra lattice plane ends a strong distortion of the crystal lattice is obtained. The line that extends along the end of the extra half-plane is called a dislocation line.

Dislocations cannot end inside the crystal. They always pass through the crystal up to surface or grain boundary or they form dislocation loops inside the crystal.

Crystal defects

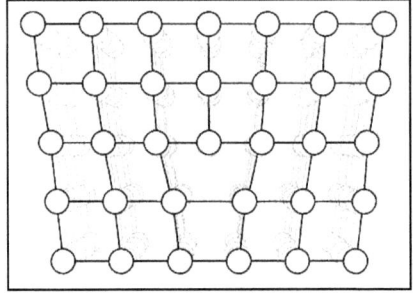

Figure 26: An extra half plane of atoms is added to the crystal lattice forming an edge dislocation.
Reference: IFW Dresden

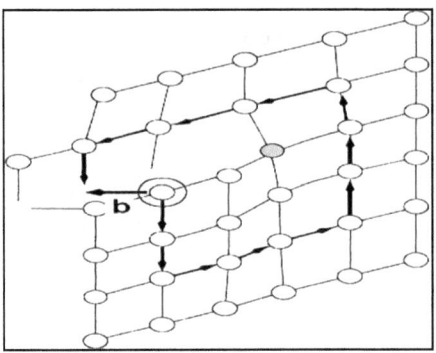

Figure 27: A screw dislocation.
One part of the crystal lattice is shifted relative to the other crystal part. Atomic planes form a spiral surface around the dislocation line. b is the Burger's vector. *Reference: IFW Dresden*

A dislocation can be characterized by its dislocation line and Burger's vector. A Burger's circuit surrounds a perfect, defect free region of a crystal. Starting point and end of this circuit are the same (figure 28). When the Burger's circuit surrounds a dislocation, the starting and end points are not the same (figure 28). The difference between the two points is the Burger's vector $\vec{b}$.

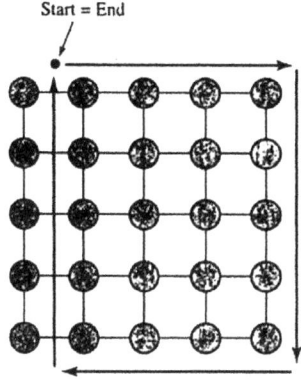 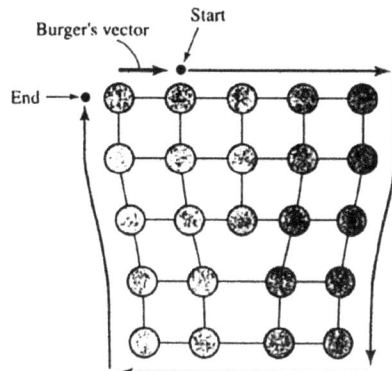

Figure 28: Illustration of Burger's circuit and Burger's vector.
Reference: *http://www.mrl.uscb.edu*

As mentioned in Chapter 2 the incorporation of dislocations in the growing crystal can be avoided by using the Dash technique.

3.3 Stacking faults

Stacking faults are two-dimensional area defects inside the crystal lattice. The face centred cubic structure of silicon can be described as a stacking of crystallographic planes with the following sequence: ABCABC... . If an additional plane is added to the crystal lattice the original sequence is altered - ABACABC instead of ABCABC for example. Stacking faults which are caused by the insertion of an extra plane are called extrinsic stacking faults.

If a crystal plane is removed from the stack the original sequence is interrupted (ABCBCABC...). Stacking faults obtained in this way are called intrinsic stacking faults. The extrinsic stacking faults found in silicon crystals are normally caused by condensation of silicon interstitials. These can be created in the oxidation of silicon wafers or due to precipitation of impurities like oxygen during annealing processes [18]. Stacking faults can be decorated by metals like copper or iron which enhance their electrical activity resulting in degraded device performance.

3.4 Point defects and their agglomerates

3.4.1 Introduction

Intrinsic point defects are always incorporated into the growing silicon crystal. Voronkov found that the type of grown-in defects, interstitials or vacancies, is dependent on the ratio of the pulling speed V to the axial temperature gradient G between the growing crystal and the melt [16].

The critical V/G value ξ_T is about 0.13 mm^2/min.K. Crystals grown with a higher V/G are vacancy rich. If the V/G value is lower than ξ_T, interstitials dominate. This simple but very important V/G rule holds both for Float-zone and Czochralski grown crystals [16]. When the crystal cools down point defects can agglomerate into larger defects.

3.4.2 Swirl defects (A- and B-defects)

If a crystal is grown with a ratio V/G < ξ_T the predominance of self interstitials is favoured and the vacancy concentration is undersaturated [16]. Self interstitials agglomerate into swirl defects. Two different kinds of swirl defects are known: larger ones which are called A-defects and smaller ones which are called B-defects [16].

The typical size of A-type swirl defects is around one micron in FZ material and a few microns in Czochralski (CZ) grown crystals [19].

Swirl defects were first discovered in dislocation-free Float Zone (FZ) material by preferential etching and X-ray topography after copper decoration [20] (chapter 4). Their nature was lateron studied by high-voltage transmission electron microscopy (HVEM) by Kolbesen and Föll.

The FZ grown silicon wafers were etched with a modified "Sirtl Solution" and then thinned to a residual thickness of ~ 2µm [21]. After Sirtl etching two different kinds of etch structures, namely small shallow etch pits and etch hillocks, were obtained. The etch hillocks were caused by A-swirl-defects and the small etch pits were probably caused by B-swirl-defects. The etch hillocks were characterized in detail by high-voltage transmission electron microscopy (HVEM). In most cases dislocation loops or complicated arrangements of dislocation loops were found below the hillocks. The density and the size of the etch hillocks are dependent on the concentration of impurities like carbon [21].

Figure 29 shows two examples of single loops found in silicon crystals with a relatively high impurity content. Figure 30 shows clusters of dislocation loops. It is possible to identify the single dislocation loops. Complicated arrangements of loops were found in FZ crystals grown under vacuum with a low impurity content (figure 31). In this case it was not possible to distinguish between the individual loops [21].

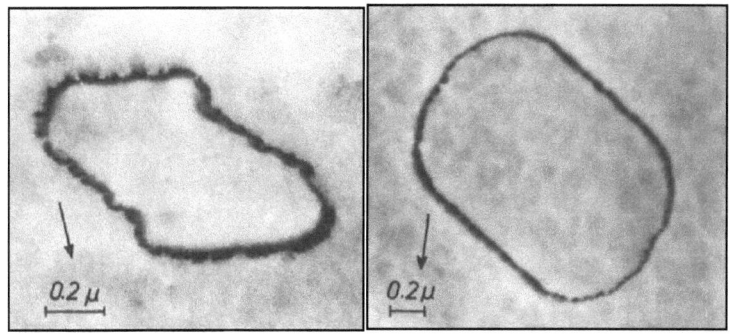

Figure 29: Single dislocation loops found below the "Swirl hillocks" after etching.
Reference: *H. Föll and B. O. Kolbesen Applied Physics (8) 1975*

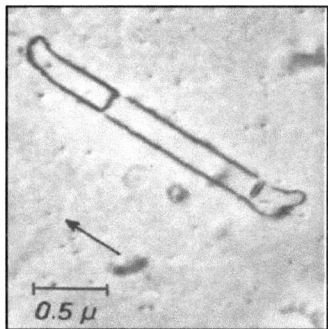

Figure 30: Simple arrangement of dislocation loops. It is possible to distinguish the individual loops.
Reference: H. Föll and B. O. Kolbesen Applied Physics (8) 1975

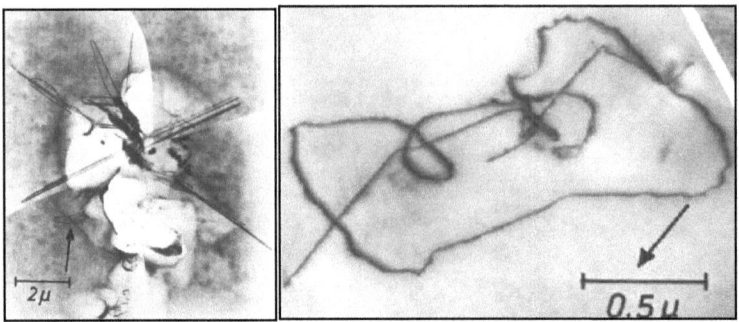

Figure 31: Complex arrangement of dislocation loops. It is not possible to distinguish between the individual loops.
Reference: *H. Föll and B. O. Kolbesen Applied Physics (8) 1975*

Swirl loops always show an anomalous contrast which may be ascribed to decoration effects. The dislocation loops can be decorated by impurities like carbon or oxygen and metals for example. It is also known that swirl-loops can generate very small dislocation loops themselves around the "main defect" These small dislocation loops appear as dark spots (figure 32) [21].

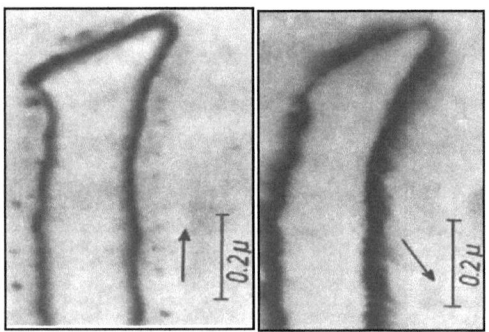

Figure 32: HVEM images of decorated dislocation loops.
The black dots and the hem-like contrast may have been caused by small dislocation loops.
Reference: *H. Föll and B. O. Kolbesen Applied Physics (8) 1975*

B-swirl-defects are not well characterized due to their weak strain field [3].The B-swirl-defects are not small dislocation loops but loosely packed three dimensional agglomerates of self interstitials and some impurity atoms like carbon. They are less stable than A-swirl-defects [3]. The B-Swirl-defects could be interpreted as forerunners of the A-Swirls [21].

3.4.3 D-defects (vacancy agglomerates or COPs)

If a crystal is grown with a V/G ratio higher than ξ_T, the intrinsic point defects that result are mainly vacancies. The interstitial concentration is undersaturated. The interstitials recombine with vacancies. When the crystal cools the excess vacancies agglomerate into voids (D-defects or COPs) [16].

Voids or D-defects are small (50-200 nm) and octahedral (figure 33). They appear often as twins (figure 33).

The octahedral shape of the voids was found by Takemi et al. using a copper decoration technique followed by thinning of the decorated samples by the focussed-ion-beam technique. Later on they used a new method for the delineation of the vacancy agglomerates. The three-dimensional coordinates of the grown-in defects were first determined by infrared laser scattering tomography. The samples were then thinned by the focused-ion-beam technique for detailed TEM investigations. In Czochralski grown material a thin (~ 2nm thick) oxygen layer was found inside the voids by energy-dispersive x-ray spectroscopy (EDS) and Auger electron spectroscopy. The oxygen layer inside a cavity is caused by the formation of SiO_x species [22].

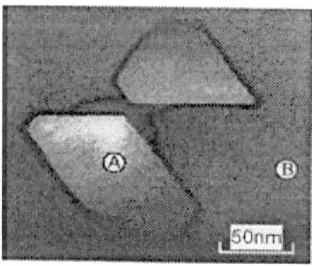

Figure 33: Cross sectional TEM image of a D-defect. The octahedral twin structure can be recognized.
The spots A and B are measurement points for EDS and AES investigations. Reference: *T. Ueki Appl. Phys. Letters 70 (10), 1997*

The density of voids is low ($10^6 - 10^7$ cm^{-3}). It is dependent on the cooling rate and the local vacancy concentration according to the equation given by Falster.

$$N_V = \frac{B\,(-dT/dt)^{3/2}}{C_V^{1/2}}$$

where

N_V = the final void concentration around the nucleation temperature T_N
B = is a constant dependent on vacancy and void parameters
$(-dT/dt)$ = the cooling rate at reaction temperature
C_V = the local vacancy concentration and is a function of V/G.

An efficient formation of voids occurs within a short temperature interval during the cooling process. This "nucleation temperature", T_N, lies between 1000°C and 1100°C [19]. When the void concentration becomes too high the vacancy population decreases fast and further void formation is suppressed [23]. In the case of CZ grown silicon crystals the formation of oxide particles is now favoured. These oxide particles are formed by aggregation of vacancies and oxygen. The vacancy zone is subdivided into a major inner part containing voids and a marginal band (P-band) containing mainly oxide particles [24]. In the case of FZ grown silicon crystals the stopped formation of voids results in residual vacancies and small vacancy clusters which are "frozen in". The void density obtained by simulation is about 3×10^7 cm^{-3} while the density of the small vacancy clusters is much higher ($\sim 10^{12}$ cm^{-3}) [19]. Voids (or D-defects) can cause problems in the manufacture of high density ICs and dynamic random-access memory (DRAM) chips [16]. Leakage currents which lead to an increased power consumption and heating are caused by D-defects for example. In DRAM chips a void can cause a short circuit between two memory cells.

3.4.4 Variation of V and G during the crystal growth process

The first small diameter dislocation free silicon crystals were grown with a relatively high G value which led to the predominance of interstitials. Large diameter crystals produced today are grown with a lower value of G, resulting in the formation of vacancy agglomerates (D-defects) [19].

It is possible to grow silicon crystals with a changing V/G ratio. The variation of the pulling speed V and/or the axial temperature gradient G leads to the formation of different coexisting regions in the silicon crystal. A gradual increase of the pulling speed V during the crystal growth process leads to the formation of an interstitial rich region at the edge of the crystal and a vacancy rich area in the centre. These two regions are separated by a denuded zone with a very low point defect concentration [19].

3.4.5 Oxygen precipitates

Float Zone (FZ) and Czochralski (CZ) grown silicon crystals contain oxygen. In FZ material the oxygen content is about 10^{16} atoms/cm^3. Owing to the nature of the process the oxygen content in the CZ material is higher being around 10^{18} atoms/cm^3 (chapter 2). The oxygen atoms are incorporated as interstitials into the crystal lattice (figure 34).

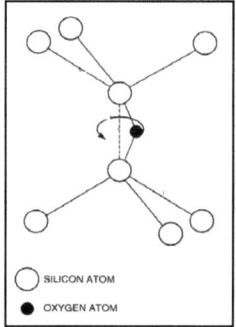

Figure 34: Oxygen interstitial atom (Oi) incorporated in the crystal lattice.
Reference: *R. C. Newman J. Phys. Condens. Matter (12), 2000*

An interstitial oxygen atom O_i is located between two silicon atoms. The Si-O_i-Si arrangement is angular (interbonding angle: 160°, figure 35) [25]. In the silicon melt at ~ 1420°C the solubility of oxygen is about 10^{19} O_i / cm^3, decreasing with decreasing temperature. At 1000°C the solubility of oxygen inside the silicon crystal is reduced to ~ 10^{17} O_i / cm^3. That means that only a small part of the oxygen is solved in the silicon crystal homogeneously. The main part of the oxygen forms a solid SiO_x-phase. [7]. Together with the accompanying vacancies oxygen atoms could form octahedral particles [19]. Without vacancies the formation of these oxide particles is suppressed [19]. The formation of silicon dioxide (SiO_2) precipitates in the silicon crystal is not possible due to the large volume misfit between Si and SiO_2. SiO_2 precipitates would produce a strong strain field in the crystal lattice. Only after high temperature annealing (T > 1100°C), are amorphous SiO_2 precipitates formed in CZ grown material [25]. It was calculated by Voronkov and Falster that the combination of a vacancy and two oxygen atoms generate an unstrained oxide particle.

$\gamma = (\eta-1)/2$

where

γ = the vacancy/oxygen ratio which is ~ 0.5. (the vacancy to oxygen ratio of ~ 0.5) ratio as a relation

η = the oxide/silicon volume ratio.

As mentioned before an efficient formation of voids occurs within a short temperature interval around the nucleation temperature T_N. If the number of voids produced becomes too high the local vacancy concentration C_V decreases and the further formation of voids is suppressed.

At a lower vacancy concentration the formation of oxide particles is favoured. The oxide particles exist mostly as vacancy-O_2 species [19]. If a crystal contains both, a vacancy- and an interstitial rich area, then C_V decreases to zero at the vacancy/interstitial border [23]. The vacancy zone is then subdivided into a zone with a relatively high vacancy concentration C_V (where the voids are formed) and an area with a low vacancy content where the oxide particles are formed [23]. This marginal oxide particle band (P-band) runs along the vacancy/interstitial border and also along the crystal surface [23]. The vacancy-O_2 complexes are too small to be detected. After annealing these small oxygen nuclei agglomerate to a detectable size [26]. The defect density of the larger oxide particles can be determined by infrared laser scattering tomography (IR-LST). The size and shape of these oxide particles is dependent on annealing temperature and annealing time. Detailed investigations were performed by Sueko and co-workers. They heated separate CZ-grown wafers in a N_2 atmosphere at 700, 800 and 900°C for 64, 200, 400 and 700h at each temperature. The oxygen content was about 1.6×10^{18} atoms/cm^3. After heat treatment the samples were thinned mechanically and by argon sputtering to prepare them for TEM investigations.

Differently shaped oxygen precipitates, dependent on annealing time and temperature, were found by weak-beam TEM investigations (table 5) [27].

Temperature (°C)	Annealing time (h)	Shape of the oxide particles
700	64	Sphere/Plate
700	200	Plate
700	400	Plate
700	700	Plate
800	64	Plate
800	200	Plate
800	400	Plate/Polyhedra
800	700	Polyhedra
900	64	Plate
900	200	Plate
900	400	Plate
900	700	Polyhedra

Table 5: Morphology of the oxygen precipitates found by weak-beam TEM investigations.

3.4.6 Oxidation-induced stacking faults (OSF) and OSF ring

High temperature annealing (T > 1150°C) of CZ grown wafers in an oxygen atmosphere leads to the formation of an OSF ring. The OSF ring appears as an annular ring with a high density of oxidation-induced stacking faults [28] and can be revealed by preferential etching. Oxygen precipitates were found close to the centre of these stacking faults [29]. The ring radius R_{OSF} is dependent on the crystal growth rate V and the temperature gradient G [28]. Small oxygen precipitates inside the P-Band act as nuclei for OSF. The oxidation of silicon leads to formation of self-interstitials at the Si/SiO_2 interface. These excess interstitial atoms agglomerate at the oxygen nuclei and can be revealed by preferential etching (figure 35). They are called oxidation-induced stacking faults (OSF or OISF). The growth of the OSF starts with the beginning of the thermal process. For that reason all OSF have practically the same size (figure 36).

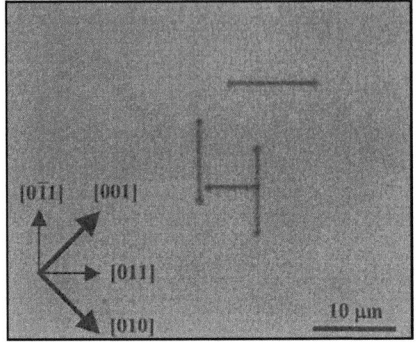

Figure 35: Oxidation-induced stacking faults, revealed by Secco etching.
Reference: *H Soyama and H. Kumano Electrochemical and Solid State Letters 3 (2), 2000*

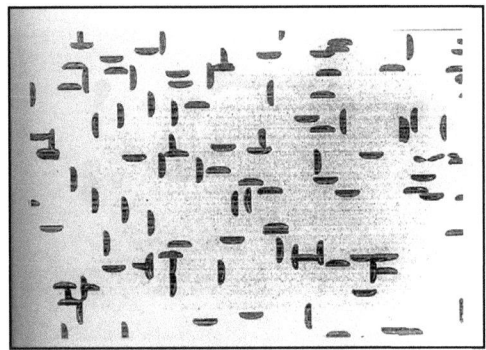

Figure 36: Surface OSF on CZ material. Found after etching with Secco solution (3 min.). Magnification: 200 x. Note that all OSF have the same length. *Reference: H. Rauh Wacker's Atlas for Characterization of Defects in Silicon*

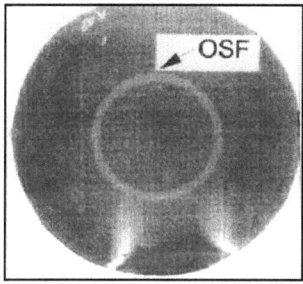

Figure 37: X-ray topograph of an OSF ring. The CZ wafer was first heated to 1200°C in a wet O$_2$ atmosphere and then etched.
Reference: Yamagishi et al Semicond. Sci. Technol. 7, 1992

If the pulling speed V is higher then 1 mm/min. the OSF region disappears and voids (D-defects) are found distributed all over the wafer [30].

3.5 Perfect (nearly defect free) silicon crystals

There are three different ways to produce perfect, nearly defect free silicon crystals.

3.5.1 Epitaxy-process:

A new crystal layer can be built up at the surface of a wafer by chemical vapour deposition (VPD, chapter 2.5). This new "epi-layer" is nearly defect free - it may contain only a few stacking faults.

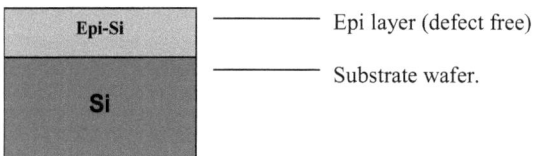

Figure 38: "Nearly defect free" epitaxial layer grown on a substrate wafer.

3.5.2 Crystal growth process:

Crystals grown with V/G < ξ_T are interstitial rich and those grown with V/G > ξ_T are vacancy rich. It is possible to produce crystals with a very low point defect concentration if the V/G ratio lies around the critical value ξ_T [16]. Growth processes which can control V/G to within 10% of the critical value ξ_T axially and radially are capable of producing such nearly defect free silicon [16].

3.5.3 Argon annealing:

Another method of fabricating defect free silicon is to anneal silicon wafers at about 1100°C in an argon atmosphere [31]. At this high temperature the thin oxide film inside the voids is destroyed. Oxygen diffuses out and interstitials recombine with vacancies forming a perfect crystal lattice [31]. This method produces a perfect silicon crystal up to a depth of 10 µm [31]. Defects, especially voids, located in deeper parts of the wafer are desired because they act as gettering sinks for impurities like iron or copper [31].

4 Delineation of crystal defects by chemical etching

4.1 Introduction

Certain crystal defects in silicon wafers like dislocations, stacking faults, Swirl defects or vacancy agglomerates can be delineated by chemically etching the surface. They are revealed as pits or hillocks on the surface and can be characterized and their frequency of occurrence determined under an optical light microscope. This is a well-established method for quality control [10]. For a more detailed characterisation of the etched figures transmission electron microscopy (TEM), scanning electron microscopy (SEM) or atomic force microscopy (AFM) can also be used.

Generally etching solutions consist of:

- an oxidizing agent - normally a Cr^{6+} species or nitric acid – to convert Si into SiO_2
- a fluoride component - normally HF - which dissolves the SiO_2
- a diluent, usually water or acetic acid
- certain additives like bromine, iodine or Cu^{2+}

The etching process consists of two steps. At first silicon is oxidized:

$Si + Ox \rightarrow SiO_2$
Ox = oxidizing agent

The silicon dioxide formed is then dissolved by hydrofluoric acid:

$SiO_2 + 4 HF \rightarrow SiF_4 + 2 H_2O$
$SiF_4 + 2 HF \rightarrow H_2 [SiF_6]$

The formation of SiO_2 at the silicon surface is rather unlikely. It is only a formal step. An alternative mechanism for the oxidation of silicon in HF/HNO_3 mixtures is described by Steinert et al., and will be discussed later on. When the steady-state method (d[intermediates]/dt • 0) is applied it becomes obviously that the oxidation of silicon is the rate-determining step.

By repeating the process of oxidation and dissolution the surface layer can be progressively removed. The removal can be demonstrated easily by covering one part of a wafer or wafer fragment with wax. This part will not be attacked during the etching process. After etching the wax layer is removed with organic solvents. The difference in height between the etched and the protected areas, which can be determined using a profilometer corresponds to the layer removed. In the case of SOI material the thickness of the layer removed can be determined by ellipsometry as the difference between the initial thickness of the SOI layer and the layer thickness after etching. The removal rate can be calculated by dividing this difference by etching time.

$$r = \frac{\Delta d}{t_{etch}} = \frac{\Delta m}{t_{etch}}$$

r = removal rate
Δd = removed silicon thickness
Δm = mass loss
t_{etch} = etching time

The removal rate (r) should be dependent on the concentration of the oxidizing agent and the HF content.

$$r = k \, [Ox]^a \, [HF]^b$$

k = rate coefficient for the etching process dependending on etching solution, material and temperature.
a, b = partial orders of reaction.

At a given temperature the removal rate is a characteristic parameter of the etching solution.

4.2 General aspects of etching silicon

4.2.1 Diffusion-controlled etching mechanism

As mentioned before, the mechanism of etching in silicon can be divided into two different elementary processes namely the oxidation of silicon and the dissolution by hydrofluoric acid of the silicon dioxide formed. These two steps can be diffusion or reaction-controlled. In the case of a diffusion-controlled mechanism the rate coefficient is dependent on the diffusion coefficients of the reactants:

$$\frac{d[SiO_2]}{dt} = k_D\,[Ox] \quad \text{and} \quad \frac{d[SiF_4]}{dt} = k_D\,[SiO_2]\,[HF]$$

$k_D = 4\pi R D N_A$
and
k_D = rate coefficient for a diffusion-controlled mechanism
D = diffusion coefficient of the corresponding reactant
R^* = distance between the reactive species and crystal surface
N_A = Avogadro constant = 6.022×10^{23} mol^{-1}

The diffusion coefficient D itself is dependent on the hydrodynamic radii R_{ox} and R_{HF} of the oxidizing agent and the hydrofluoric acid respectively and the viscosity η of the etching solution used.

$$D = \frac{kT}{6\pi\eta R_{Ox}} \quad \text{and} \quad \frac{kT}{6\pi\eta R_{HF}}$$

When $R^* = \sim R_{ox}$ or R_{HF}, then:

$$k_D = \frac{8RT}{3\eta}$$

An increase in temperature corresponds to a decrease in viscosity of the etching solution and an increase in the mobility of the different reactive species. The decrease in viscosity is given by the Arrhenius equation [32]:

$$\eta = A\, e^{\frac{E_{vis}}{RT}}$$

η = viscosity of the etching solution
A = pre-exponential factor
E_{vis} = activation energy for the elementary process of the viscous shift
R = gas-law constant = 8.414 JK^{-1}mol^{-1}
T = temperature (K)

The diffusion coefficient D and the reaction rate increase with decreasing viscosity.

4.2.2 Reaction-controlled etching mechanism

In the case of a reaction-controlled mechanism the reaction rate depends on how fast the reactants absorb the required energy from ambient molecules.
The removal rate r increases with increasing temperature following Arrhenius law. At a higher temperature more molecules have the energy necessary to form the activated complex.

$$r = A\, e^{\frac{-E_a}{RT}}$$

r = removal rate
A = pre-exponential factor
E_a = activation energy of the complete etching process
R = gas-law constant = 8.414 JK^{-1}mol^{-1}
T = temperature (K)

4.2.3 Characterisation of the etching mechanism by its activation energy E_a

The activation energy is a characteristic parameter of the etching solution. It varies with etching solution and material etched but is independent of temperature. To determine the activation energy of an etching process it is necessary to determine the removal rates at different temperatures. The activation energy can be calculated from the slope, if ln r is plotted against 1/T.

$\Rightarrow \ln r = \ln A - E_a/RT$

An activation energy of around 15 kJ/mol or less is characteristic of a diffusion-controlled etching mechanism. Activation energies of around 40 kJ/mol or more are characteristic of reaction-controlled etching mechanisms [32].

4.3 Classification of etching solutions

There are two different classes of etching solutions:

- Polishing etches → mainly diffusion-controlled
- Structural etches → mainly reaction-controlled

Structural etches can be divided into preferential and non-preferential etches

Polishing etches

Polishing etches have a levelling effect on the crystal surface. Particles which adsorb at a crystal surface are attacked preferentially due to their higher potential energy. If the surface is planar and smooth a homogeneous removal is obtained (figure 39). Polishing etches are normally not able to delineate crystalline defects. Etched figures are not formed [32]. High removal rates and a low activation energy of the etching process are typical for polishing etches.

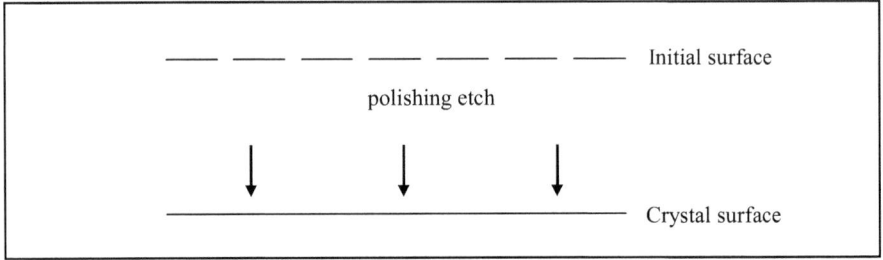

Figure 39: Equal attack of a polishing etch at a crystal surface.

Structural etches

Structural etches can be divided into preferential and non-preferential etches. Both produce etched figures, normally etch pits, at crystal surface. A preferential attack at crystal imperfections like dislocations, stacking faults or vacancy agglomerates leads to the formation of etch pits due to an increased removal rate (figure 42 and 43).

Defects in a crystal may cause mechanical tensions inside the crystal lattice. These tensions result in a lowered activation energy for the etching process (figure 40 and 41). A decrease in activation energy corresponds to an increase in removal rate.

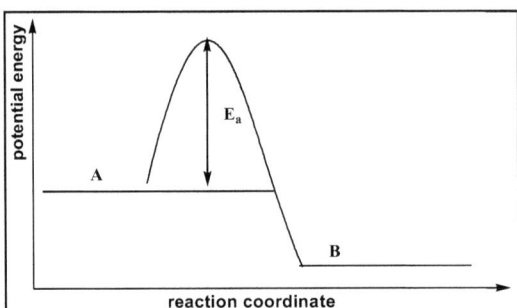

Figure 40: Activation energy for the etching process at a perfect crystal lattice. EA = activation energy for etching process. A = initial state, B = final state.

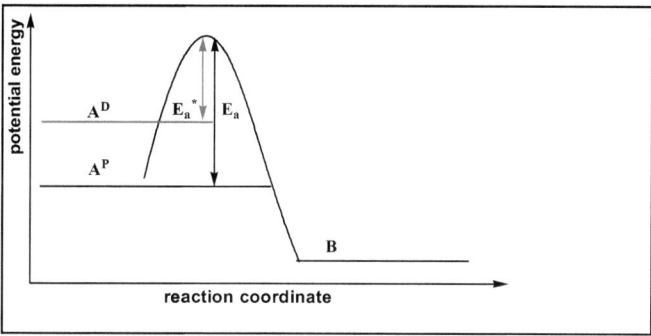

Figure 41: Schematic representation of the difference in activation energies between a perfect lattice, and a defect site. A^P = initial state (perfect crystal lattice)
A^D = initial state (dislocation)
B = final state
E_A = activation energy for the etching process in a perfect crystal.
E_A^* = reduced activation energy for the etching process at a crystal defect.

In non-preferential etches, the removal rate is made up of three different removal rates (C_1, C_2 and C_3):

C_1 = removal rate on a perfect crystal (perpendicular to the surface)

C_2 = removal rate on a perfect crystal (parallel to the surface)

C_3 = removal rate at a crystal defect ($C_3 > C_1$, C_2)

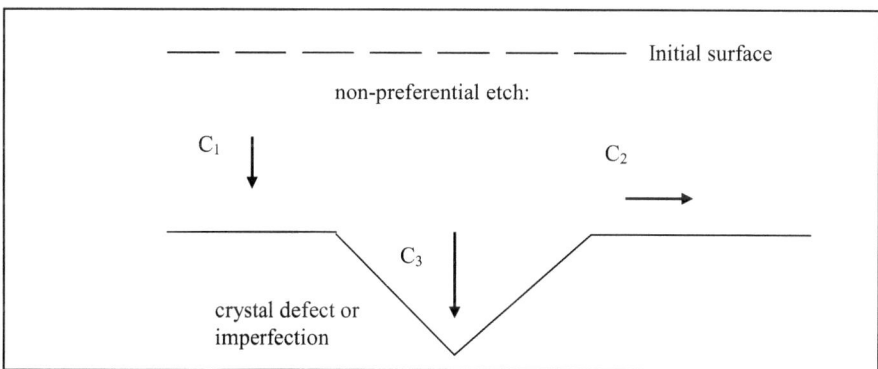

Figure 42: The removal rates of a non-preferential etching solution.

Non-preferential etching solutions normally produce circular or oval-shaped, shallow etch pits.

In preferential etches the removal rate consists of four different partial removal rates (C_1 - C_4). At a crystal defect the removal rate is increased which is similar to non-preferential etches but the various crystal faces are also attacked differently (figure 43).

C_1 = removal rate at the perfect crystal perpendicular to the surface
C_2 = removal rate at the perfect crystal parallel to the surface
C_3 = removal rate at crystal defect ($C_3 > C_4$)
C_4 = removal rate at a crystal face ($C_4 > C_1, C_2$)

Figures formed by preferential etches are characteristic of the etching solution used they can be circular, triangular, square or oval shaped.

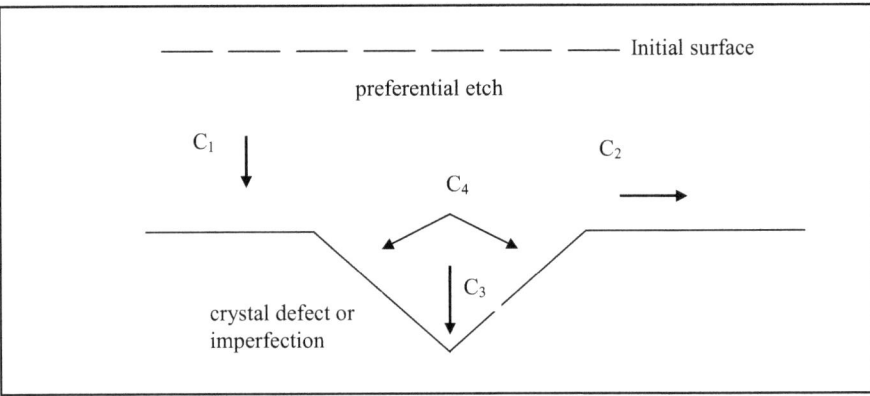

Figure 43: The removal rates of a preferential etch (C1 - C4).

Etch hillocks are normally caused by , precipitates, usually metals or agglomeration of self-interstitials. The agglomeration of additional atoms or ions lowers the removal rate compared to that of the perfect silicon crystal. Polishing and structural etches are borderline cases. Depending on the composition of the etching solution, concentration of the reaction products and temperature a polishing etch can change into a structural etch or vice versa. The material etched and the kind of crystal defect can turn non-preferential etches into preferential etches. The composition of an etching solution alone does not determine if it is structural or polishing. The terms structural etch and preferential etch are normally used synonymous.

4.4 Chromium based etch recipes

4.4.1 The Secco Solution

Most of the structural etches used today contain a Cr^{6+}-species, normally $K_2Cr_2O_7$ or CrO_3, as the oxidizing agent. The most frequently used structural etch is the Secco solution. It is able to reveal different kinds of defects like vacancy agglomerates, stacking faults or dislocations in {100} or {111} oriented, p- or n-doped silicon with a resistivity ranging from 4 to 300 Ωcm. The Secco solution does not work on highly doped material. Diluted versions can be used for the delineation of defects in thin and very thin films [10].

Instead of $K_2Cr_2O_7$, $NaCr_2O_7$, $NH_4Cr_2O_7$ or CrO_3 can also be used. During the etching process the colour of the solution can change from orange to brown-green. This effect is caused by the formation of Cr^{3+} species [33]. The Secco solution shows a linear dependence of the removal on etching time. The removal rate originally published by Secco et al. lies around 1.5 µm/minute. It can be increased and the formation of bubbles, which can cause artefacts at the crystal surface, can be decreased by using ultrasonic agitation during the etching process [33]. The removal rates of the different Secco formulations, shown on table 5, were determined at room temperature (25°C) without stirring or ultrasonic agitation. The required amount of potassium dichromate was dissolved in DI water after which hydrofluoric acid was added to the solution. All solutions were kept in polypropylene vessels. The solutions were only used one day after preparation to guarantee a complete dissolution of the potassium dichromate.

The removal rate of the original Secco recipe was determined on 725 µm thick p-doped, {100} oriented silicon substrates with a resistivity of 25 Ωcm.

The experimentally determined removal rate differs distinctly from the published value (table 5). For the delineation of defects in SOI or sSOI material it is necessary to decrease the removal rate. It can be decreased by diluting the etching solution and hence reducing the hydrofluoric acid and the potassium dichromate content (table 6). The etching rates of the two different Secco diluted recipes were determined on standard Smart-Cut^{TM} SOI material. The exact course of the reaction regarding the oxidation process of the silicon is still not clear. Different chromium(VI)- species like $Cr_2O_7^{2-}$, $HCr_2O_7^-$ or $H_2Cr_2O_7$ are present in acid solutions. They have a high affinity to silica surfaces due to interactions between the chromate and protonated silanol groups (figure 44).

The reduction mechanism of the chromium(VI) is not known. An electron transfer from the silicon to the Cr^{6+} (figure 45) should be the most probable reaction mechanism [34].

Van den Meerakker and van Vegchel suggested the following etching mechanism:

(a) $Cr\ (VI) \rightarrow Cr\ (III) + 3\ h^+$
(b) $Si + h^+ \rightarrow Si^+$

First of all the Cr (VI) compound is reduced by injection of holes into the valance band of the silicon resulting in the formation of Si^+ species. The Si^+ species may stabilized due to the present of F^- [35]. The Si^+ species are rather unstable. Electrons are injected into the conduction band, resulting in the formation of intermediates with higher oxidation states [35]:

(c) $Si^+ \rightarrow Si\ (IV) + 3\ e^-$

The authors suggested an electron transfer to protons (H^+) in the solution which leads to the formation of hydrogen [35]:

(d) $Si^+ + 3\ H^+ \rightarrow Si\ (IV) + 3/2\ H_2$

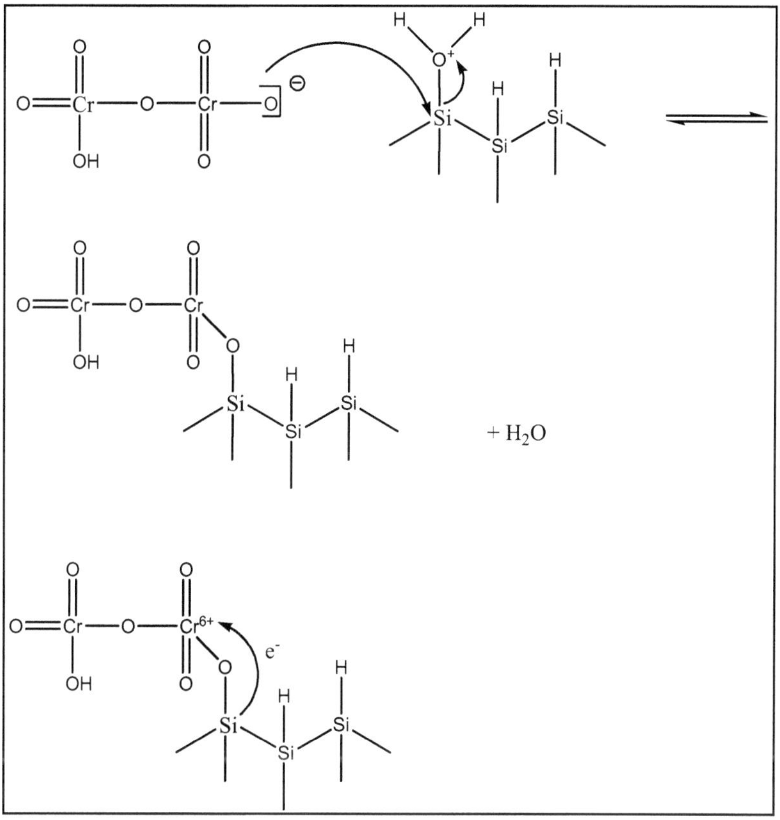

Figure 44: Possible reaction mechanism for the reduction of chromium(VI)-compounds in acid solutions. The first step is the adsorption of chromate at the silicon surface. The next step is an electron transfer from the silicon to the chromium(VI)-species.

Etching solution	Composition	Removal rate (nm/minute) at 25°C
Secco	2.2g $K_2Cr_2O_7$ / 50 ml H_2O / 100 ml HF (50%) c Cr^{6+} = 0.1 mol/l	770 ± 4 1500*
Secco dil. 1	0.6g $K_2Cr_2O_7$ / 100 ml H_2O / 50 ml HF (50%) c Cr^{6+} = 0.027 mol/l	44.4 ± 1
Secco dil. 2	0.3g $K_2Cr_2O_7$ / 100 ml H_2O / 50 ml HF (50%) c Cr^{6+} = 0.014 mol/l	31 ± 0.8

Table 6: Compositions and properties of the different Secco recipes. The removal rates were determined on silicon substrates and SOI material.

The influence of temperature on removal rate was investigated for each Secco solution. Therefore, the removal rates were determined at four different temperatures. The temperatures were adjusted by using a thermostat. The removal rate is exponentially dependent on temperature as expected (figure 45-47). The influence of temperature on the removal rate was determined for each of the Secco solutions mentioned in Table 7. The removal rate was found to increase exponentially with temperature as shown in Figures 45 and 46.

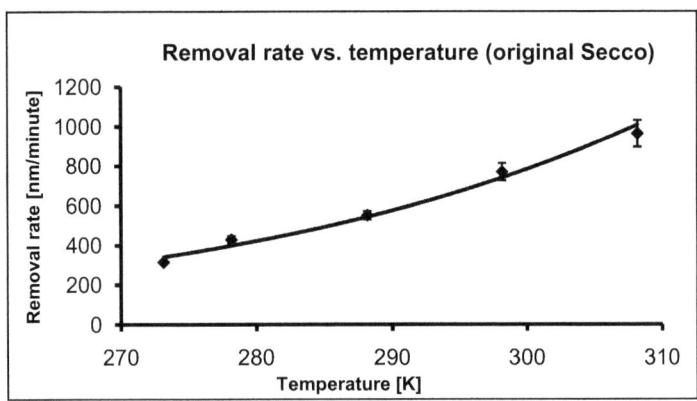

Figure 45: Removal rate vs temperature determined on standard Si-Bulk material.

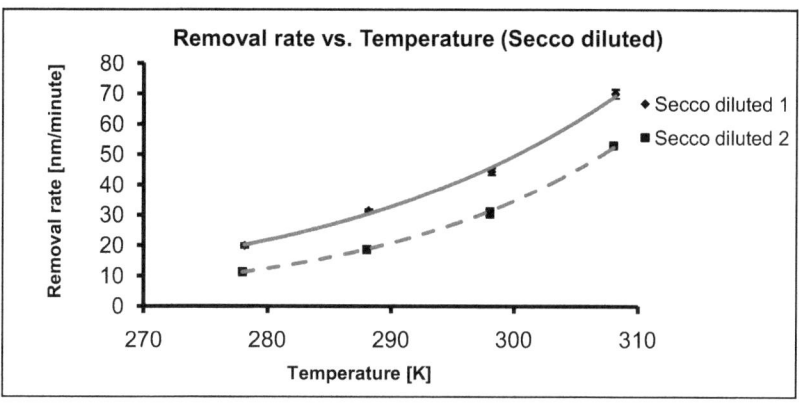

Figure 46: Temperature dependence of the removal rates shown for the Secco-diluted 1 and the Secco-diluted 2 solutions. The removal rates were estimated on SOI material.

The activation energies for the etching process of silicon were determined by plotting ln of the removal rate r against the reciprocal of the temperature (1/T) as mentioned before (figure 47 and table 7). The activation energy for the etching process is a characteristic parameter for each etching solution.

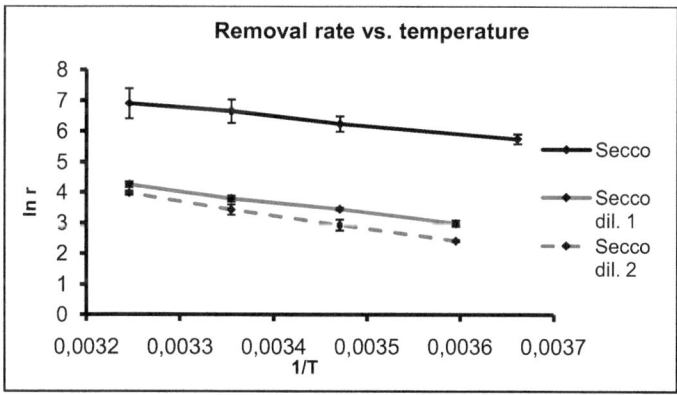

Figure 47: The activation energies of the original Secco recipe and the Secco diluted 1 and Secco diluted 2 solution. r = removal rate and T= temperature.

Etching solution	Removal rate (nm/minute at 25°C)	Activation energy E_a for the etching process (kJ/mol)
Secco	770	23
Secco diluted 1	44.4	29.3
Secco diluted 2	31	36.5

Table 7: Experimentally determined activation energies of the different Secco solutions.

In case of the Secco diluted 1 and the Secco diluted 2 solutions the experimentally determined activation energies are relatively high. They suggest a more reaction-controlled etching mechanism. A relatively low activation energy was found in the case of the original Secco solution. Therefore, a more diffusion-controlled etching mechanism should be expected. This experimental result however, stands in contrast to the capability of the Secco solution to reveal different crystalline defects in silicon substrates. A higher activation energy for the etching process should be expected.

4.4.2 Further chromium based etch recipes

The compositions of further chromium based etch recipes are shown in table 8. The Wright etch is normally used to reveal different kinds of stacking faults. It produces well-developed etched figures which can be recognized clearly. Bulk faults are revealed as loops, oxygen induced stacking faults are delineated as extended partial discs. The Wright solution can be used for {100} or {111} oriented, p- or n-doped silicon crystals [36].

The Schimmel solution is normally used for the delineation of crystal defects in Si/Ge alloys.

Etching solution	Composition	Removal rate (nm/minute) at 23°C
Sirtl	50g CrO_3 /100 ml H_2O/100 ml HF (50%) c Cr^{6+} = 2.5 mol/l	1300
Schimmel	7.5g CrO_3 / 100 ml H_2O/200 ml HF (50%) c Cr^{6+} = 0.25 mol/l	1000
Wright	45g CrO_3 /270 ml H_2O/ 6g $CuNO_3$* $3H_2O$/ 90 ml HNO_3 (69%)/180 ml HAc (100%)/ 180 ml HF (50%) c Cr^{6+} = 0.625 mol/l	1000

Table 8: Compositions of further currently used chromium based etching solutions.

4.4.3 Toxicity of chromium (VI) compounds

Chromium(VI) – species are highly toxic by all exposure ways. Chronic exposure to chromium (VI) dust or vapour can affect the respiratory system and cause ulcerations and perforations of the septum, bronchitis and a reduced lung function. The International Agency for Research on Cancer (IARC) has classified chromium (VI) as carcinogenic to humans [5]. Epidemiological studies of workers in the production of chromate or chromate pigments show a clear connection between exposure to chromium (VI) compounds and lung cancer [5]. Chromium (VI) compounds are unstable in vivo. Chromium (VI) is reduced by enzymes to chromium (V), chromium (IV) and finally to the less toxic chromium (III). It is believed that the toxicity of chromium (VI) compounds lies in its ability to damage cellular components during metabolism, this metabolic process,[5]. Potassium dichromate ($K_2Cr_2O_7$) may also be toxic to the reproductive system and the developing foetus [5]. For this reasons the usage of Cr (VI) compounds will be restricted worldwide and there is an urgent need for environmentally friendly chromium-free etching solutions.

4.5 Existing chromium free recipes

Only a few chromium free recipes which can be used for the delineation of crystal defects in silicon substrates and new materials like SOI or sSOI are known and used today. These solutions consist generally of HNO_3, HF and HAc (Table 9). There are not many chromium-free recipes for the delineation of crystal defects in silicon substrates and new materials like SOI and sSOI.

Etching solution	Composition	Removal rate (nm/minute)
Dash	43 ml HNO_3 (69%) / 14.5 ml HF (50%) / 143 ml HAc (100%) HF : HNO_3 : HAc 1 3 10	2.2 (25°C)
FS Cr-free SOI	95 ml HNO_3 (69%) / 12 ml HF (50%) / 93 ml HAc (100%) HF : HNO_3 : HAc 5.9 47.7 46.4 with 0.5 ml Br_2 added to 100 ml etching solution!	3.5 (23°C)
Jeita	78 ml HNO_3 (69%) / 6 ml HF (50%) / 36 ml HAc (100%) / 36 ml H_2O HF : HNO_3 : HAc : H_2O 1 13 6 6	321 (25°C) ± 19
MEMC	104 ml HNO_3 (69%) / 8 ml HF (50%) / 24 ml HAc (100%) / 45,6 ml H_2O HF : HNO_3 : HAc : H_2O 1 13 3 5.7	469 (25°C) ± 28

Table 9: Chromium free etching solutions based on the HF/HNO_3/HAc system

The Dash solution is able to reveal dislocations and stacking faults in silicon substrates, but produces etched figures that are poorly developed and have blurred outlines. The etch removal with Dash is not uniform. In SOI and sSOI samples etched with Dash delamination of the thin silicon film on the SiO_2 occurred which makes it unsuitable for the delineation of defects in thin and very thin films and its low removal rate at room temperature limits its use on silicon substrates.

The FS Cr-free etch was developed for the delineation of defects in standard Smart-Cut^{TM} SOI material with SOI layers that are less than 100nm thick. Etch removal is uniform and etch pits are well formed [37].

Jeita and MEMC are the most frequently used chromium-free etch recipes. They are able to reveal D-defects, dislocations or stacking faults and have uniform etch removal, the thickness of the removed layer increasing as a function of etching time. However, the removal rates (etching rates), which increase exponentially with temperature are much too high for the delineation of defects in SOI and sSOI (see Table 9). Removal rates of Jeita and MEMC solutions were determined at four different temperatures on 725 µm thick, p-doped, {100} oriented silicon with a resistivity of 25 Ωcm. The activation energy for the etching process (of silicon) was derived from the plot in Fig. 48.

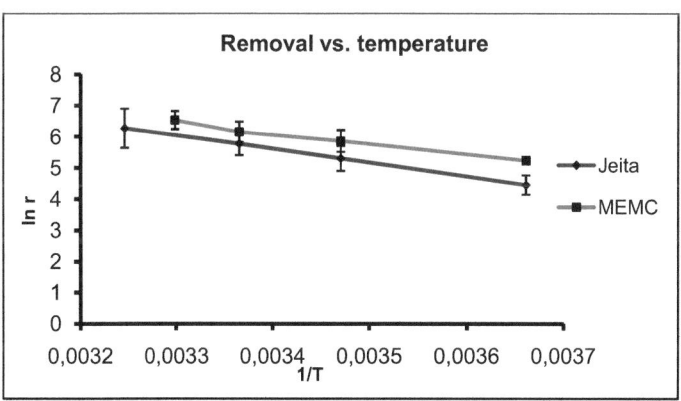

Figure 48: Removal rate as a function of temperature.

Etching solution	Removal rate (nm/minute at 25°C)	Activation energy E_a for the etching process (kJ/mol)
MEMC	469	28.6
Jeita	321	36.5

Table 10: Experimentally determined activation energies of the MEMC and Jeita solution.

The activation energies found for the Jeita and MEMC solutions are in the same range as those of the Secco diluted recipes. The relatively high activation energies indicate a more reaction controlled etching mechanism.

4.6 Chemistry of HNO_3 / HF mixtures

The etching of silicon in HNO_3/HF mixtures is often described as a process that consists of three different steps: the formal oxidation of silicon to SiO_2 by nitric acid (1); the dissolution of the SiO_2 by HF (2) and finally the formation of $H_2[SiF_6]$. The overall reaction (4) suggests that nitrogen monoxide, hexafluorosilicic acid and water are the only reaction products [38].

(17) $3\ Si + 4\ HNO_3 \rightarrow 3\ SiO_2 + 4\ NO + 2\ H_2O$
(18) $SiO_2 + 4\ HF \rightarrow SiF_4 + 2\ H_2O$
(19) $SiF_4 + 2\ HF \rightarrow H_2[SiF_6]$

(20) $3\ Si + 4\ HNO_3 + 18\ HF \rightarrow 3\ H_2[SiF_6] + 4\ NO + 8\ H_2O$

The rate limiting step is the oxidation of silicon. Abel et al. postulated the formation of nitrous acid during the etching process (equation 22). The nitrous acid should be the actual oxidizing agent and is assumed to be generated autocatalytically during the etching process itself:

(21) $HNO_2 + HNO_3 \rightleftharpoons N_2O_4 + H_2O$
(22) $N_2O_4 + 2\ NO + 2\ H_2O \rightleftharpoons 4\ HNO_2$

According to Robbins et al. the formation of HNO_2 is dependent on the nitrous and nitric acid content [39].

$$\frac{d[HNO_2]}{dt} = k\,[HNO_3]\,[HNO_2]$$

The HNO_2 is assumed to be the active species which oxidizes the silicon [39]:

(23) $4\ HNO_2 + Si \rightleftharpoons SiO_2 + 2\ H_2O + 4\ NO$

Freshly prepared etching solutions (of HNO_3/HF) have an induction period with significantly lower etching rates. The induction period can be avoided by adding a nitrite salt to the solution [39]. The results of such experiments confirm that the nitrous acid is the reactive species.

Kelly at al. proposed an alternative reaction path in which the nitrosonium ion (NO^+) acts as the actual oxidizing agent [40]. To explain the etching mechanism of silicon in HNO_3/HF mixtures new detailed investigations were performed by Steinert et al. At first the dependence of the removal rate on nitrous acid content was investigated. Different amounts of a nitrite salt were added to etching solutions containing equal volumes of hydrofluoric acid and nitric acid. It was found that the removal rate increased in direct proportion to the nitrite content. These results confirmed the reaction pathway suggested by Robbins et al. The role of intermediate species which appear during the etching process was also investigated. Silicon fragments were etched in a freshly prepared HNO_3/HF mixture without the addition of nitrite ions at 1°C. The concentrations of nitrite, nitrate and fluoride ions were determined by ion chromatography. After the etching process had started the colour of the solution changed from light-brown to deep blue. The blue colour is caused by the formation of N_2O_3. The presence of N_2O_3 was confirmed by UV-vis and Raman spectra [41]. Further (NIII)-species, namely a complex ion ($3\ NO^+\ NO_3^-$) also denoted as $N_4O_6^+$ and the nitrosonium ion NO^+ were also found. These results confirm the work of Kelly et al. who suggested the formation of NO^+ in HNO_3/HF mixtures. Other nitrogen species like NO_2^+ were not found. The presence of nitrite ions could also be ruled out with Raman spectroscopy [41]. The nitrosonium ion (NO^+) is assumed to be the reactive species which oxidizes the silicon. Steinert and co-workers suggested that the N_2O_3 acts as a reservoir for the formation of NO^+:

(24) $N_2O_3 + H^+ \rightleftharpoons HN_2O_3^+$

(25) $N_2O_3 + H_2O \rightleftharpoons 2\ HNO_2$

(26) $HNO_2 + H^+ \rightleftharpoons H_2NO_2^+$

(27) $H_2NO_2^+ + H^+ \rightleftharpoons H_3O^+ + NO^+$

For the ion chromatography the etching solutions have to be diluted. During the dilution process the reactive intermediates undergo a decay which leads to the formation of nitrite ions [41].

The nitrite concentration measured in diluted HF/HNO$_3$ mixtures by ion chromatography, acts as a sum parameter for the reactive N(III)-species in concentrated solutions [38]:

$[NO_2^-]_{\text{dil. soln.}} = [N_2O_3]_{\text{conc. soln.}} + [NO^+]_{\text{conc. solution}} + [HN_2O_3^+]_{\text{conc. sol.}} + [H_2NO_2^+]_{\text{conc. sol.}} + [HNO_2]_{\text{conc. sol.}}$

In the case of HF-rich systems which contain hydrofluoric acid in the range of 90 to 60 percent by volume the etching rate is dependent on the concentration of the reactive N(III) intermediates. The activation energies for the etching process range from 41 kJ/mol to 44 kJ/mol, depending on the concentration of the N(III) species. The removal rate decreases on increased stirring of the etching solution because the reactive N(III) intermediates are removed from the silicon surface. These experimental results indicate a reaction controlled etching mechanism of silicon in HF-rich systems.

In the case of HNO$_3$-rich mixtures the etching rate is almost independent of the concentration of the reactive N(III) species. Increased stirring during the etching process leads to a higher removal rate. The activation energies of the etching process range from 17 kJ/mol to 34 kJ/mol depending on the content of the N(III) intermediates. At high N(III) concentrations the reaction mechanism is diffusion-controlled. At low N(III) concentrations the etching process is „more likely, mechanism-controlled [38].

The experimentally determined activation energies for the SEH and MEMC solutions are in the same range as those found by Steinert et al.for HF/HNO$_3$ mixtures with a low content of reactive N(III) species.

Etching mechanism of silicon in HF/HNO$_3$ systems

The silicon surfaces of etched wafer fragments were investigated by X-ray-induced photoelectron spectroscopy. A Si-H terminated surface was always found. The absence of

Si-O species even after etching of silicon fragments in HNO_3-rich mixtures indicates that silicon dioxide is not formed. Steinert et al. suggested a divalent electrochemical dissolution of the silicon [41]. The etching process is initiated by nitric acid which generates holes (h^+) due to an oxidation process. These holes are formally located at the silicon surface (figure 49). The next step is a nucleophilic attack of HF or HF_2^- species at the positively charged silicon surface which leads to the formation of Si-F bonds (figure 50).
Finally the polarized Si-Si bonds are attacked by HF or HF_2^- forming SiF_4 (figure 51). The SiF_4 reacts with excess hydrofluoric acid forming H_2SiF_6 [41].

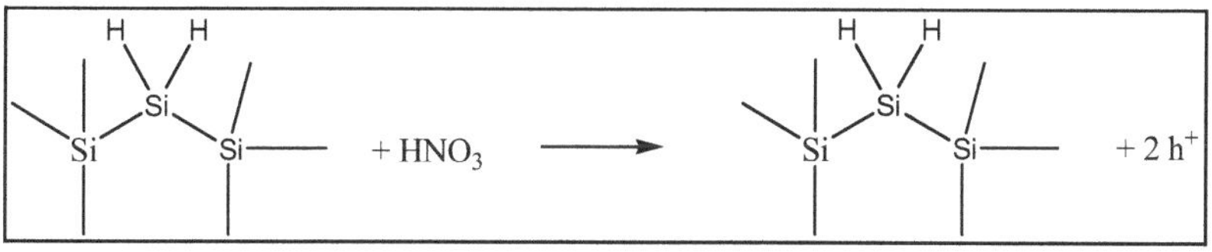

Figure 49: Etching mechanism of silicon in HF/HNO_3 mixtures after Steinert et al. The first step is the partial oxidation of silicon which leads to the formation of holes (h+).

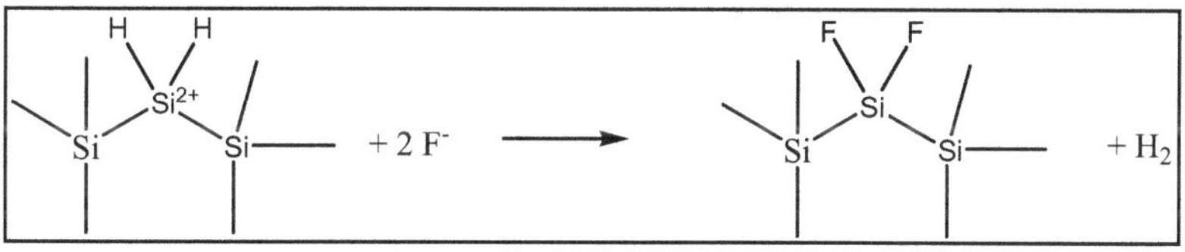

Figure 50: The second step is a nucleophilic attack of F- ions which results in the formation of Si-F bonds.

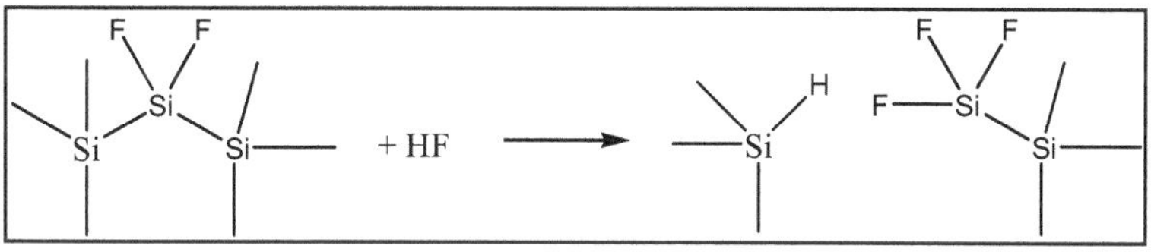

Figure 51: The polarized Si-Si bonds are attacked by HF.

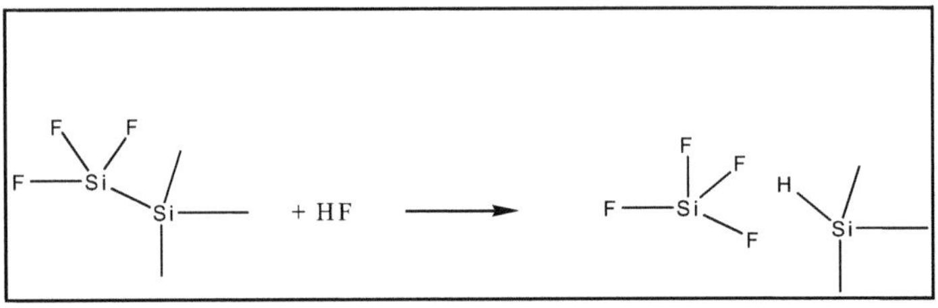

Figure 52: The last step is the formation of SiF$_4$

5 Organic Peracid Etches: A new class of chromium free etching solutions for the delineation of defects in different silicon based materials

5.1 Introduction

Chromium-based etching formulas are very efficient in revealing different kinds of crystal defects in various semiconducting materials. Chromium (VI) compounds, however, are highly toxic and carcinogenic and in the near future their use will be heavily restricted worldwide.

Only a few chromium-free etching solutions are in use today. Most of these solutions have one or more disadvantages. Some are only able to reveal a certain class of crystal defects, others do not work on thin and very thin films. There is a demand for environmentally friendly etching solutions which can also be used for new materials like SOI or sSOI.

The subject of this study is a new class of environmentally friendly chromium-free etching solutions based on peracids that was developed to delineate the various crystal defects found in thin films and silicon substrates.

These new etching solutions will be referred to as Organic Peracid Etches (OPE) and consist of a short-chain alkanoic acid - normally acetic acid - hydrogen peroxide and hydrofluoric acid. A few hours after putting them together the corresponding peracid is formed which probably acts as the oxidizing agent [10]:

$$(28) \quad R-\overset{O}{\underset{O-H}{\bigg\langle}} \; + \; H_2O_2 \; \rightleftharpoons \; R-\overset{O}{\underset{O-O-H}{\bigg\langle}} \; + \; H_2O$$

$R = H, -CH_3, -CH_2CH_3$ or $CH_2-CH_2-CH_3$

Instead of the alkanoic acid the corresponding acid anhydride or chloride can also be used to generate peracids. Short-chain peracids are frequently used as disinfectants. They have an intense, sharp odour and irritate the skin and mucous membranes, but they are less toxic than chromium (VI) compounds and can easily be destroyed by reduction [42]. H_2O_2/HF/HAc mixtures are normally used as selective Si/Ge etches to remove the SiGe transfer layer in sSOI fabrication. Such solutions have high removal rates (40 – 50 nm/min) on Si/Ge alloys but very low removal rates on silicon (0.4 – 2 nm/min. at 25°C). The basic idea is to use such solutions for the delineation of defects in silicon crystals. The lower the removal rate, the higher the selectivity should be. Organic Peracid Etches are able to reveal various crystal defects like stacking faults, dislocations, Swirl-defects and vacancy agglomerates in silicon crystals. The low removal rates make them also suitable for defect delineation in thin and very thin (<50 nm) films. At first, the Dash etchant which is the best known of the chromium-free etchants was used as a basis for developing the Organic Peracid Etches. The original Dash formulation contains nitric acid as oxidizing agent. The nitric acid was replaced by an equal volume of hydrogen peroxide:

Original Dash solution: HNO_3 (65%)   HF (49%)   Hac (100%)
                                             43 ml   15 ml   142 ml

Modified Dash solution: H_2O_2 (30%)   HF (49%)   Hac (100%)
("Organic peracid etch (OPE) A")   43 ml   15 ml   142 ml

This new solution was labelled OPE A. It had a very low removal rate (0.6 nm/min) on silicon material and was able to reveal different crystal defects. The characteristics of the original and modified Dash solutions" are summarized in table 10.

Etching solution used	Removal rate (nm/min., 25°C)	c oxidizing agent (mol/l)	c HF (mol/l)	Ratio Oxidizing agent/HF
Original Dash solution	2.2	3	2.1	1.4
Modified Dash solution	0.6	$c_{peracid}$: 1.6 c_{total}: 2.1	2.1	0.8 1

Table 11: Properties of the original and the modified Dash solution.

The removal rate of the OPE A was however found to be too low for practical use other than for the delineation of defects in thin films. Therefore the OPE A was modified in several ways to increase the removal rate and to investigate the influence of the different components on the removal.

5.2 The H_2O_2/HF/HAc system

5.2.1 Introduction

Mixtures of H_2O_2/HAc always contain peracetic acid. The formation of peracetic acid (PAA) is an example of a typically balanced reaction. The PAA concentration should be dependent on the hydrogen peroxide and the acetic acid content. D´Ans and Frey suggested a bimolecular mechanism for the formation of PAA:

$$\frac{d[PAA]}{dt} = k\,[H_2O_2]\,[H_3C\text{-}COOH]$$

The reaction should be first-order regarding the hydrogen peroxide and the acetic acid content [43]. The formation of peracetic acid can be accelerated by adding a catalyst to the H_2O_2/HAc mixture. Inorganic mineral acids like sulphuric, phosphoric or hydrofluoric acid can be used. D´Ans and Frey found out that the rate of formation of the peracetic acid is directly proportional to the amount of catalyst added to the solution [43].

The rate of formation can also be increased by increasing the temperature [44].

Pure PAA which is obtained by fractional distillation, fractional freezing and centrifuging is a colourless liquid with a characteristic sharp odour. It melts at 0.1°C, boils at ~ 25°C and is readily soluble in water and organic solvents like diethyl ether or ethyl acetate [42]. Aqueous solutions of peracetic acid can be stored a few weeks at room temperature. PAA decomposes slowly. The main reaction products are water, carbon dioxide, carbon monoxide, oxygen and acetic acid [42].

Several methods like amperometry, conductometry, photometric measurements, as well as gas chromatographic or high-performance liquid chromatographic approaches can be used to determine the PAA content in the presence of high concentrations of hydrogen peroxide. The simplest and most reliable method with which to determine the PAA content is a classical iodometric titration. The hydrogen peroxide and the peracetic acid content can be determined by adding potassium iodide to the corresponding etching solution. Both compounds are able to convert iodide into iodine:

$$\overset{-I}{H_2O_2} + 2\overset{-I}{I^-} + 2H^+ \rightarrow \overset{0}{I_2} + 2\overset{-II}{H_2O}$$

$$\underset{-II\,-I\,-I}{H_3C\text{-}COOOH} + 2\,\underset{-I}{I^-} + 2\underset{0}{H^+} \rightarrow \underset{-II\,-II}{I_2} + \underset{-II}{H_3COOH} + H_2O$$

The iodine concentration can be determined by titration with a thiosulphate solution:

$$I_2 + 2\,S_2O_3^{2-} \rightarrow 2\,I^- + S_4O_6^{2-}$$

$$c\,H_2O_2 + c\,PAA) = c\,I_2 = \tfrac{1}{2}\,c\,S_4O_6^{2-}$$

It is also possible to determine only the PAA content by adding a small amount of catalase to the buffered (pH 7) solution. The catalase accelerates the decomposition of hydrogen peroxide into water and oxygen. After a few minutes the hydrogen peroxide is completely destroyed. The amount of iodine now liberated corresponds to the peracetic acid content [44].

$$c\,PAA = c\,I_2 = \tfrac{1}{2}\,c\,S_4O_6^{2-}$$

5.2.2 Experimental procedure

For all etching experiments wafer fragments of approximately one square cm were used. One or two wafer fragments were placed vertically into a holder which was then immersed in the etching solution for a given amount of time. The wafer pieces were then rinsed with DI-water and dried under a flow of nitrogen. Each batch of the etching solution consisted of 125 -200ml and was prepared one day before use. The etching took place in polypropylene vessels, which, because of the relatively high vapour pressure of short-chain alkanoic peracids, were kept closed.

All experiments were performed at room temperature taken to be 20°- 25°C. Temperatures deviating from this range were adjusted using a thermostat.

The defect density was determined as follows: defects were counted under the microscope at 10 random sites on each wafer fragment. This was repeated on 6 - 8 wafer pieces for each experiment giving a total of 60 – 80 sites from which the average per square cm was calculated. The defect densities were determined at magnifications of 500 (area TV screen: 35200.6 cm²) or 1000 times (area TV screen: 8800.9 cm²).

The thickness of the layer removed by etching was determined by measuring the layer before and after etching with an ellipsometer and is sometimes referred to in the following simply as "the removal".

5.2.3 Influence of the hydrofluoric acid content on etching behaviour

To investigate the influence of the hydrofluoric acid content on etching behaviour different etching solutions with a constant peracetic acid content and varying HF concentrations were prepared (table 12). Generally all solutions were used one day after preparation to allow the peracid to develop fully. The peracid concentration was determined by iodometry. The removal rate of each etching solution was determined by etching several SOI fragments at room temperature (20°C).

Six wafer pieces were used for each etching solution. The thickness of the SOI layer was measured with an ellipsometer before and after etching the difference being the thickness of the layer removed. All values shown in table 12 are mean values.

Etching solution used	Composition H_2O_2 (30%)	HF (50%)	HAc (100%)	c peracetic acid (mol/l)	c HF (mol/l)	Removal rate (nm/minute)
OPE 10	43 ml	4 ml	153 ml	1.54	0.58	0.36
OPE 11	43 ml	8 ml	149 ml	1.5	1.16	0.36
OPE A	43 ml	15 ml	142 ml	1.64	2.1	0.32
OPE 20	43 ml	22 ml	135 ml	1.52	3.19	0.34
OPE 21	43 ml	29 ml	128 ml	1.6	4.2	0.34
OPE B	43 ml	43 ml	114 ml	1.6	6.24	0.32

Table 12: Composition of organic peracid etches with varying HF concentrations and the removal rates.

The HF content was increased progressively from 2 vol. % up to approximately 25 vol. %. The peracetic acid content was always around 1.6 mol/l. The removal rate was nearly the same for all etching solutions. Within this range of hydrofluoric acid concentrations the removal rate of H_2O_2/HF/HAc is not influenced by the hydrofluoric acid concentration. All mixtures shown on table 11 are able to reveal crystal defects in the SOI-layer.

5.2.4 Influence of the concentration of hydrogen peroxide and peracetic acid on removal rate

The oxidation of silicon is the rate-limiting step in the etching process. The removal rate is only dependent on the concentration of the oxidizing agent. Organic Peracid Etches always contain two possible oxidizing agents, namely hydrogen peroxide and the peracid. Both compounds should be able to convert silicon into silicon dioxide:

(29) $Si + 2\ H_2O_2 \rightarrow SiO_2 + 2\ H_2O$

(30) $Si + 2\ R\text{-COOOH} \rightarrow SiO_2 + 2\ H_3C\text{-COOH}$

where R = H (performic acid); $-CH_3$ (peracetic acid); $-CH_2-CH_3$ (perpropanoic acid); or $-CH_2-CH_2-CH_3$ (perbutyric acid)

The corresponding short-chain peracids are stronger oxidizing agents than hydrogen peroxide. Silicon is not etched if the alkanoic acid is replaced by an equal volume of water (table 13). Therefore it is assumed that the peracid is the reactive species which oxidizes the silicon.

Etching solution used	Composition	Material used	T (°C)	Etching time (minutes)	Removal (nm)	Removal rate (nm/minute)
TS 1	50 ml H_2O_2 (30%) 50 ml HF (50%) 100 ml DI-H_2O	Smart-Cut™ SOI Initial thickness: 88 nm	25	80	0	0
TS 2	50 ml H_2O_2 (50%) 50 ml HF (50%) 100 ml DI-H_2O	Smart-Cut™ SOI Initial thickness: 88 nm	25	80	3	0.04

Table 13: Effect on etching behaviour of replacing the alkanoic acid with water

To investigate the influence of the peracetic acid content on removal rate different etching solutions with increasing peracetic acid (PAA) content were prepared.

5.2.5 Peracetic acid content as a function of time

The formation of peracids is a balanced reaction. To investigate the influence of the PAA concentration on removal rate the peracetic acid must have attained its maximum concentration in the mixture. Therefore the formation of peracetic acid as a function of time was determined for Organic Peracid Etches (OPE) with varying H_2O_2/HAc/HF contents. All experiments were performed at 21°C. Aqueous solutions of hydrogen peroxide with a content of 30 mass percentage (c H_2O_2 = 18.5 mol/l) and 50 mass percentage (c H_2O_2 = 33.3 mol/l) were always used. The highly concentrated H_2O_2 solution contains stabilizers in the range < 1 ppm. The PAA and the hydrogen peroxide content were determined by iodometry. The concentration of hydrogen peroxide plus PAA was determined as follows:

500 µl of each etching solution was made up to 250 ml with DI water. From this an aliquot of 25 ml was taken and an excess of potassium iodide, 2ml of 7 N sulphuric acid and 300 µl of a saturated solution of ammonium heptamolybdate added. The solution was then titrated against a 0.1 M thiosulphate solution to determine the concentration of iodine released. Ammonium molybdate acts as a catalyst in the oxidation of iodine. When potassium iodide is added the solution changes from colourless to yellow due to the formation of I_3^-.

The liberated iodine reacts with the excess iodide forming I_3^-:

$$I_2 + I^- \rightleftharpoons I_3^-$$

At the equivalence point the conversion of the liberated iodine to iodide is complete and the yellow colour disappears.

To determine the PAA concentration the solution was first buffered with a $H_2PO_4^-/HPO_4^{2-}$ mixture. 10 µl of catalase was added and the mixture allowed to stand for 5 minutes for the hydrogen peroxide to break down completely. The solution was then titrated following the procedure for the iodometric titration described for hydrogen peroxide plus PAA.

The PAA and the hydrogen peroxide content were determined in triplicate every 1-2 hours until the peracid concentration no longer changed. When the peracetic acid concentration reached a maximum the equilibrium constant K was determined from the measured and calculated values:

$$K = \frac{[PAA][H_2O]}{[HAc][H_2O_2]}$$

The total water concentration was calculated from the water contents of the hydrofluoric acid and the hydrogen peroxide solutions and the amount of water which is set free during the oxidation of the acetic acid (equation 28):

c H_2O $_{(total)}$ = c H_2O $_{(HF)}$ + c H_2O $_{(Hydrogen\ peroxide)}$ + c H_2O $_{(reaction)}$

With the following H_2O_2/HAc/HF mixtures the effect of the composition on removal rate was studied:

OPE A: H_2O_2 (30%)   HF (50%)   HAc (100%)
         43 ml   15 ml   142 ml

OPE 2: H_2O_2 (30%)   HF (50%)   HAc (100%)
         73 ml   15 ml   112 ml

OPE B: H_2O_2 (30%)   HF (50%)   HAc (100%)
         43 ml   43 ml   114 ml

OPE C: H_2O_2 (50%)   HF (50%)   HAc (100%)
         50 ml   50 ml   100 ml

OPE 24: H_2O_2 (50%)   HF (50%)   HAc (100%)
         50 ml   25 ml   125 ml

OPE F: H_2O_2 (50%)   HF (50%)   HAc (100%)
         75 ml   5 ml   120 ml

The HF concentrations, maximum PAA contents and equilibrium constants K for each etching solution are summarized in table 14.

Etching solution	c HF (mol/l)	c_{max} PAA (mol/l)	c_{max} reached after	K (21°C)
OPE A	2.11	1.64 ± 0.02	~ 20 h	4.34 ± 0.05
OPE 2	2.11	2.08 ± 0.07	~ 72 h	3.45 ± 0.12
OPE B	6.24	1.61 ± 0.005	~ 8 h	6.9 ± 0.03
OPE C	7.26	2.93 ± 0.03	~ 10 h	6.65 ± 0.07
OPE 24	3.63	3.08 ± 0.03	~ 24 h	4.6 ± 0.05
OPE F	0.73	3.67 ± 0.06	~ 72 h	3.1 ± 0.05

Table 14: Properties of different H_2O_2/HAc/HF mixtures. The maximum PAA content is obtained after 8-72 h, dependent on the composition of the etching solution.

Figure 53 and 54 show the time-dependence of the hydrogen peroxide concentration and the PAA content for the OPE A and B. In both solutions the maximum peracetic acid concentration was ~ 1.6 mol/l. In the case of the OPE A which has a relatively low HF content the maximum PAA content was attained after 20 hours (figure 53). In the case of the OPE B which has a higher HF content the maximum PAA concentration was attained after only 8 hours (figure 54) and the rate of formation of the peracetic acid was ~ 2.5 times higher compared to that of the OPE A. These investigations confirm earlier results from D´Ans and Frey.

The rate of formation of peracetic acid is dependent on the hydrofluoric acid concentration. The higher the HF content the faster the equilibrium concentration is reached (table 14, figure 53 and 54).

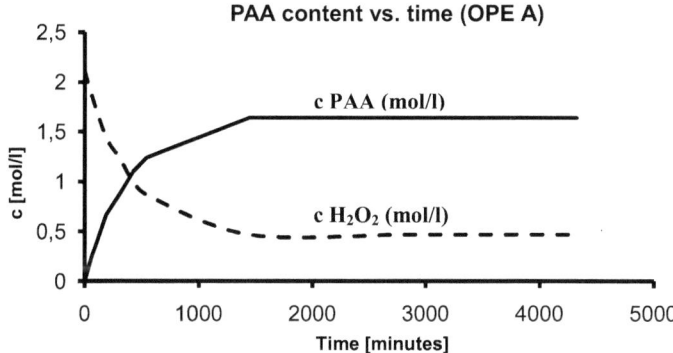

Figure 53: PAA content as a function of time determined for OPE A.

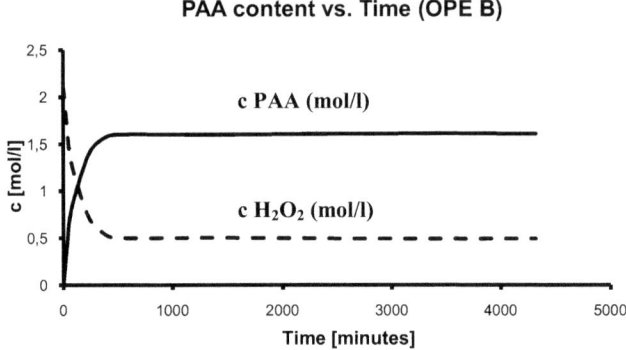

Figure 54: PAA content of OPE B as a function of time.

As mentioned before aqueous solutions containing peracetic acid are relatively stable. The peracetic acid decomposes slowly. Figure 55 shows the decomposition of the PAA with time in the OPE C. The solution was kept at room temperature and the PAA content was determined by iodometry. A notable decay of the PAA only set in after two weeks.

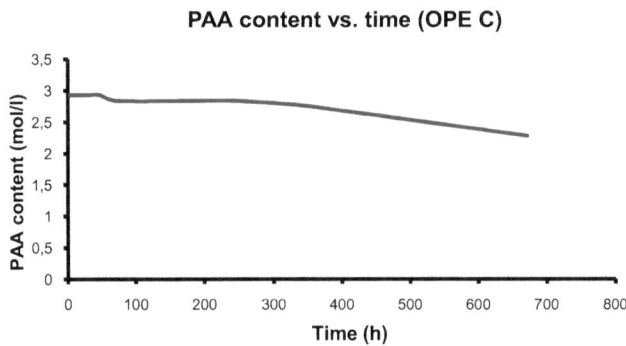

Figure 55: PAA content of OPE C as a function of time.

5.2.6 Dependence of removal rate on PAA content

Different H_2O_2/HAc/HF mixtures were prepared to investigate the influence of the PAA content on removal rate (table 14). All solutions were used after the maximum PAA content was attained (table 14). The removal rates were determined at 25°C. Six SOI fragments were used for each etching solution.

The thickness of the SOI layer was determined before and after etching with an ellipsometer. The removal rate is almost directly proportional to the peracetic acid concentration as shown on table 15 and figure 56.

Etching solution used	Composition			c PAA (mol/l)	Removal rate (nm/minute)
OPE 5	H_2O_2 (30%) 30 ml	HF (50%) 15 ml	HAc (100%) 155 ml	1.17 ± 0.03	0.27 ± 0.04
OPE A	H_2O_2 (30%) 43 ml	HF (50%) 15 ml	HAc (100%) 142 ml	1.64 ± 0.02	0.6 ± 0.03
OPE 2	H_2O_2 (30%) 73 ml	HF (50%) 15 ml	HAc (100%) 112 ml	2.04 ± 0.07	1.08 ± 0.09
OPE C	H_2O_2 (50%) 50 ml	HF (50%) 50 ml	HAc (100%) 100 ml	2.93 ± 0.03	1.35 ± 0.13
OPE 24	H_2O_2 (50%) 50 ml	HF (50%) 25 ml	HAc (100%) 125 ml	3 ± 0.03	1.42 ± 0.09
OPE F	H_2O_2 (50%) 75 ml	HF (50%) 5 ml	HAc (100%) 120 ml	3.67 ± 0.06	1.8 ± 0.09

Table 15: Composition and properties of different Organic Peracid Etches. The removal rate increases with increasing PAA content.

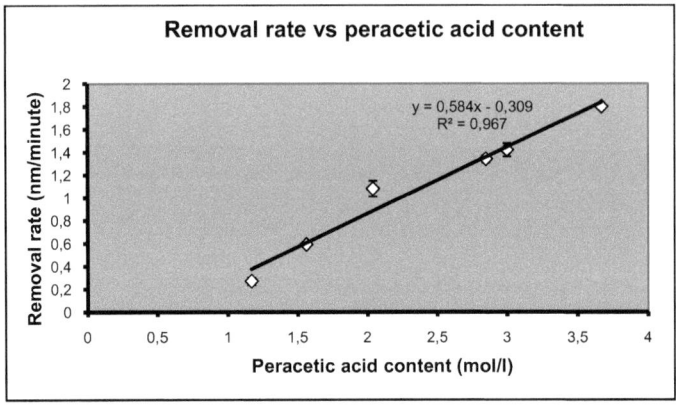

Figure 56: The removal rate as a function of PAA content.

5.2.7 Delineation of crystal defects

The etching solutions shown in Table 14 were tested for their ability to delineate defects. All the mixtures were able to reveal various types of crystal defects. Dislocations are delineated as oval-shaped etch pits (figure 57). Swirl-defects appear as loops and vacancy agglomerates (D-defects) as square shaped pits (figure 57). All solutions are also able to reveal defects in SOI material. This will be discussed later on.

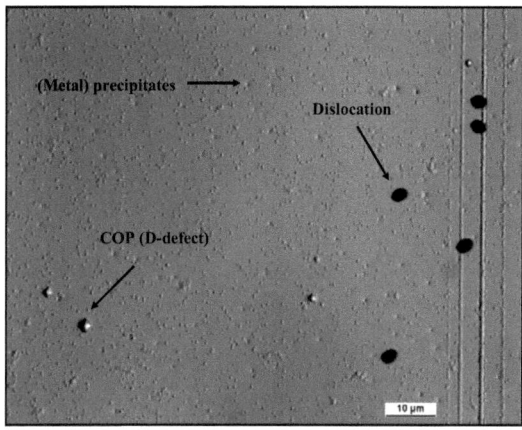

Figure 57: Optical micrograph of a CZ Si fragment after scratching and heating in a copper contaminated furnace. This fragment was etched with OPE C for 24 h at room temperature. Different kinds of defects can be recognized: dislocations (black oval shaped etch pits), vacancy agglomerates (square shaped pits) and copper decorated D-defects (small hillocks).

5.3 Etching solutions containing perpropanoic acid

The acetic acid can be replaced by propanoic acid in which case perpropanoic acid is formed instead of peracetic acid. Perpropanoic acid which is analogous to peracetic acid regarding its physical and chemical properties is more stable than peracetic acid [42].

The propanoic acid/H_2O_2/HF system has properties similar to that of the acetic acid/ H_2O_2/HF system (table 16). The perpropanoic acid is also assumed to be the reactive species which oxidizes the silicon.

Etching solution	Composition	c Peracid (mol/l)	Removal rate (nm/minute, 25°C)	K (21°C)
OPE A	H_2O_2 (30%): 43 ml HF (50%): 15 ml Acetic acid (100%): 142 ml	1.64 ± 0.02	0.6 ± 0.03	4.34 ± 0.05
OPE A.2	H_2O_2 (30%): 43 ml HF (50%): 15 ml Propanoic acid (99%): 142 ml	1.34 ± 0.02	0.64 ± 0.04	2.90 ± 0.18
OPE C	H_2O_2 (50%): 50 ml HF (50%): 50 ml Acetic acid (100%): 100 ml	2.93 ± 0.03	1.35 ± 0.13	6.65 ± 0.07
OPE D	H_2O_2 (50%): 50 ml HF (50%): 50 ml Propanoic acid (99%): 100 ml	2.59 ± 0.05	1.7 ± 0.12	6.74 ± 0.13

Table 16: Properties of the different OPE mixtures containing peracetic and perpropanoic acid. All equilibrium constants K were determined at 21°C.

The perpropanoic acid content of the OPE A.2 and the OPE D was also determined by iodometric titration and the removal rates were determined on standard Smart™-Cut SOI material at 25°C. The peracid content of the etching solutions containing propanoic acid was always lower than that of solutions containing acetic acid. Nevertheless the removal rates estimated for the OPE A.2 and OPE D are equal to or even higher than those of the corresponding OPE A and OPE C mixtures (table 16). The perpropanoic acid-containing Organic Peracid Etches are also able to reveal different defects in silicon crystals. Voids are also delineated as square-shaped etch pits (figure 58), dislocations appear as oval-shaped pits (figure 59) and Swirl defects appear as single loops or as more or less complicated arrangements of single loops (figure 60).

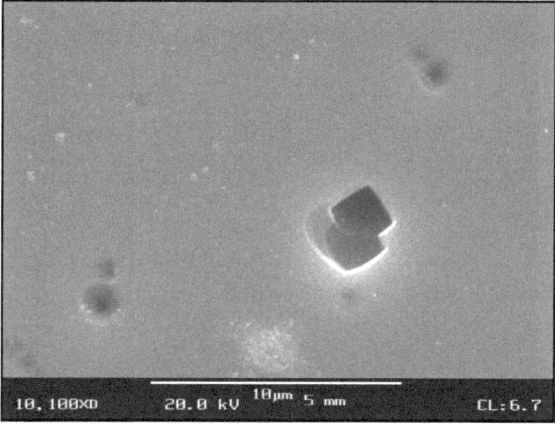

Figure 58: SEM image of a CZ Si wafer fragment after etching with OPE D. Etching time: 24 h, removal: ~ 2.5 µm. Vacancy agglomerates (D-defects) are revealed as square shaped single or double pits. The original octahedral shape of the voids can be recognized. The smaller voids below the surface have not yet been attacked by the etching solution.

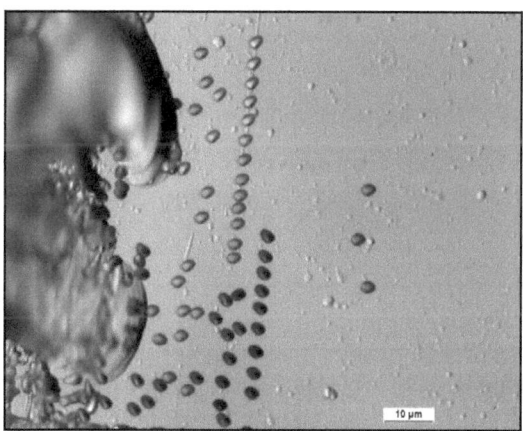

Figure 59: Wafer fragment of CZ-grown Si scratched and heated to generate dislocations. The black oval-shaped pits found after etching with OPE D (etching time: 24 h, removal: ~ 2.5 µm) are caused by dislocations.

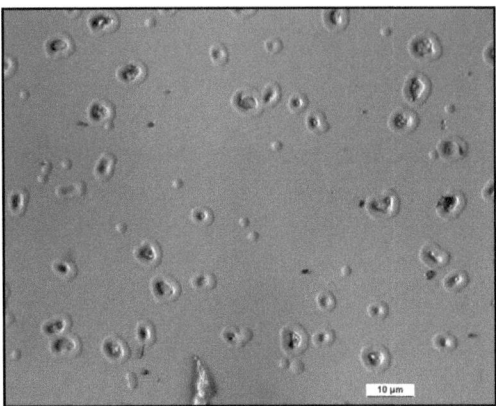

Figure 60: Swirl defects found after etching with OPE D in thin SOI material. The Swirl-defects were delineated in a small zone of silicon bulk at the edge of the SOI wafer. Etching time: 6 minutes, removal: ~ 10 nm.

5.4 Etching solutions containing perbutyric acid

Butyric acid can also be used in place of acetic acid (butyric acid/H_2O_2/HF) to produce perbutyric acid in the mixture. Perbutyric acid is similar to the other peracids discussed regarding its physical and chemical properties but is more stable than peracetic or perpropanoic acid [42]. In the mixture OPE E the acetic acid was replaced by an equal volume of butyric acid. The properties of this new solution are compared with the OPE C and OPE D in table 17. The removal rates were determined on standard Smart^{TM}-Cut SOI material at 25°C.

Etching solution	Composition	c Peracid (mol/l)	Removal rate (nm/minute, 25°C)	K (21°C)
OPE C	H_2O_2 (50%): 50 ml HF (50%): 50 ml Acetic acid (100%): 142 ml	2.93 ± 0.03	1.35 ± 0.13	6.65 ± 0.07
OPE D	H_2O_2 (50%): 50 ml HF (50%): 50 ml Propanoic acid (99%): 100 ml	2.59 ± 0.05	1.7 ± 0.12	6.74 ± 0.13
OPE E	H_2O_2 (50%): 50 ml HF (50%): 50 ml Butyric acid (99%): 100 ml	2.5 ± 0.005	1.75 ± 0.02	8.63 ± 0.1

Table 17: Composition and properties of Organic Peracid Etches containing peracetic, perpropanoic and perbutyric acid as oxidizing agents. All equilibrium constants were determined at 21°C.

The OPE E is also able to reveal different crystal defects in silicon substrates and SOI material. However, due to the unpleasant odour of butyric acid the OPE E is not suitable for practical use.

5.5 Etching solutions containing performic acid

In principle the acetic acid could also be replaced by formic acid. In this case performic acid is formed. Performic acid is the least stable of the short-chain alkanoic peracids mentioned before and decomposes rapidly into carbon dioxide and water [42]. Therefore etching solutions containing performic acid do not qualify for practical use.

5.6 Comparison of the different etch recipes

All OPE recipes were tested on SOI material and silicon substrates. Three Organic Peracid Etches (OPE C, D and F) were found suitable for practical use. These solutions produce a uniformly smooth surface after etching, have the highest removal rates and are also able to reveal various crystal defects (table 18). The OPE A and OPE B can be used for the delineation of defects in very thin films.

Etching solution	Removal rate (nm/min., 25°C)	Solution is able to reveal:			
		Dislocations	Swirl-defects	D-defects	OSF
OPE C	1.35	Yes	Yes	Yes	Yes
OPE D	1.7	Yes	Yes	Yes	Yes
OPE F	1.8	Yes	Yes	No	Yes

Table 18: Organic Peracid Etches tested on different materials.

Dislocations are revealed as oval-shaped etch pits (figure 57, 59) and vacancy agglomerates (D-defects) appear as square-shaped pits (figure 57, 58). The OPE A-D and F are also able to make A-Defects (Swirl-defects) visible. Single loops or clusters of dislocation loops are also found (figure 60). Oxydation induced stacking faults (OSF or OISF) are delineated as double pits which always have the same size (see under 5.9.3).

5.6.2 Removal vs. etching time

The influence of etching time on removal was investigated for the OPE A-D and the OPE F. Different SOI fragments with an initial thickness of the SOI layer ranging from 88 nm – 200 nm were etched. The etching time for each solution was determined according to its removal rate and the initial thickness of the SOI film. The thickness of the SOI layer was determined before and after etching with an ellipsometer. A linear dependence of the removal on etching time was found for each etching solution (figure 61).

Organic Peracid Etches

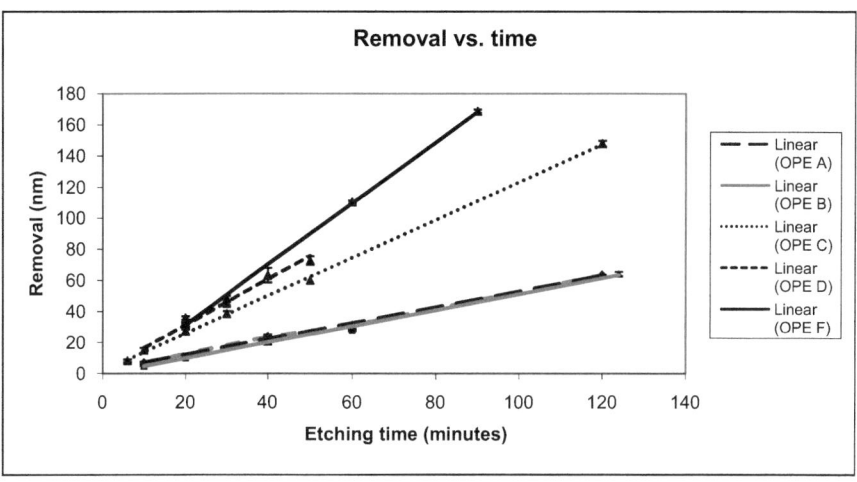

Figure 61: Removal as a function of etching time for different Organic Peracid Etches.

5.6.3 Influence of temperature

To investigate the influence of the temperature on etching behaviour SOI fragments were etched with the solutions at four different temperatures ranging from 15°C to 50°C. A smooth and plane surface was obtained at low as well as at high temperatures for each etching solution. The removal rate is exponentially dependent on temperature in accordance with Arrhenius law (figure 62).

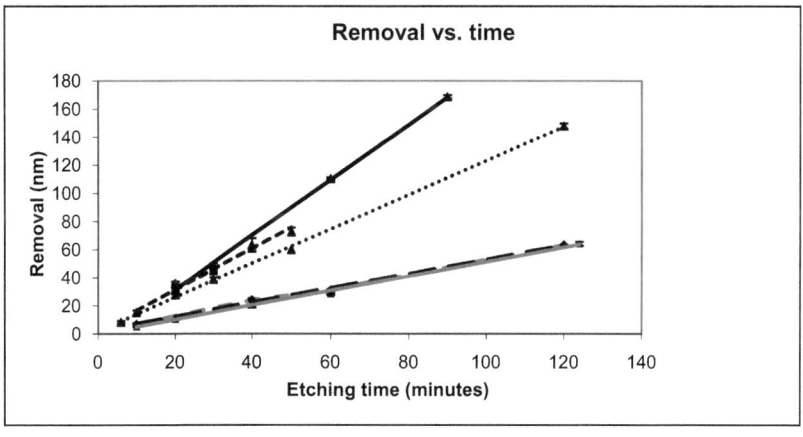

Figure 62: Exponential dependence of the removal rate on temperature.

The activation energies for the etching process of silicon were determined experimentally by plotting the ln r against 1/T. They range from 42.8 kJ/mol to 50.6 kJ/mol (table 18).

Etching solution	Removal rate (nm/min, at 25°C)	Activation energy E_a for the etching process (kJ/mol)
OPE A	0.6	47.7
OPE B	0.6	50.6
OPE C	1.3	48.7
OPE D	1.7	46.3
OPE F	1.8	42.8

Table 19: Experimentally determined activation energies for the etching process of silicon.

The high activation energies are characteristic of a reaction controlled etching mechanism. The different OPE mixtures are typical structural etches. The have the capability to reveal crystal defects with a high selectivity.

5.6.4 Influence of stirring

The experiments were performed at room temperature. SOI fragments were etched with different OPE mixtures. During the etching process the solutions were stirred at 50, 100 and 200 rpm. All experiments were performed at room temperature. It was found that neither stirring in itself nor the stirring speed had any effect on the etching behaviour of the mixtures used (supporting the assumption of a reaction-controlled mechanism). In all cases the resulting fragment surface was smooth.

5.7 Experimental results on epitaxial silicon wafers

Epitaxial silicon wafers were etched with the OPE B, C, D and F. The epitaxial-layer which is grown on a substrate wafer by chemical vapour deposition (VPD) is nearly defect-free (figure 63).

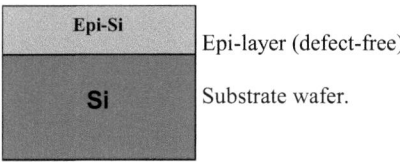

Figure 63: Nearly defect-free epitaxial layer

Wafers pieces with an epitaxial-layer thickness of about 2.1 – 2.4 µm were used. The fragments were etched with the different OPE mixtures, the duration of etching varied according to the etch rate of the mixture. A few fragments were also etched with the Secco diluted 1 solution which was used as the reference. The etching times were chosen to only partially remove the epitaxial-layer. No etch pits were expected in the residual epitaxial-layer after etching. Four different wafer pieces were etched with each etching solution, rinsed with DI-water and finally dried under a flow of nitrogen. The defects were counted under the microscope at 10 randomly chosen sites on each wafer fragment as described before. The results are summarized in table 20.

The removal was calculated from the etching time. The removal rates had been determined for each etching solution in previous experiments on SOI and Si-bulk material (see under 4.1 and 5.2). It is not possible to cover one part of the wafer fragment with wax in order to determine the removal as the difference in height after the wax layer is removed (chapter 2) because wax soils the wafer surface and produces artefacts (figure 66). It was therefore not possible to determine the defect density.

Etching solution used	Etching time	Removal (nm)	Removal (%)	Defect density cm^{-3}
Secco dil. 1	15 minutes	670	30 %	2×10^4
OPE B	12 h	400	18 %	$< 10^4$
OPE B	24 h	800	36 %	8.2×10^4
OPE C	5 h	400	18%	$< 10^4$
OPE C	10 h	800	36%	$< 10^4$
OPE C	12 h	950	42%	$< 10^4$
OPE C	20 h	1600	72%	1×10^6
OPE D	5 h	510	41%	$< 10^4$
OPE D	12 h	1260	56%	$< 10^4$
OPE F	10 h	1080	49%	$< 10^4$
OPE F	15 h	1620	75%	$< 10^4$

Table 20: Defect densities obtained on epitaxial Si with different OPE mixtures and the Secco diluted 1 as reference.
The parameters were so designed as not to remove the entire epi layer. The etching experiments were performed at room temperature (25°C). All values shown in table 19 are mean values. The removal was calculated.

At first four wafer fragments were etched with the Secco diluted reference. Approximately one third of the epitaxial-layer was removed. Small circular etch pits were occasionally found (figure 64). The defect density was very low being around 2×10^4 (table 19).

Figure 64: Wafer fragment etched with Secco diluted (reference)
Etching time: 15 min., removal ~ 670 nm, magnification: 1000 times.

In case of the OPE B small square-shaped etch pits were also found occasionally (Fig.65). They appear only after long etching times (table 19). The defect density is around 8.2×10^4 cm^{-3} which is also very low.

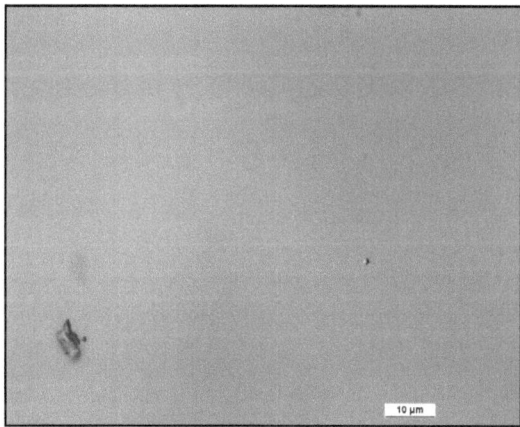

Figure 65: A single etch pit found after etching with OPE B.
Etching time: 24 h, removal: ~ 800 nm, magnification: 1000 times.

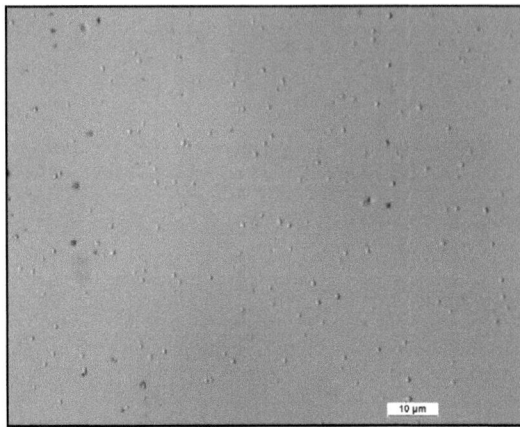

Figure 66: Optical micrograph showing a high density of etch figures found in the epitaxial-layer after etching with the OPE B for 12 h. Removal: ~ 400 nm, magnification: 1000 times. This wafer fragment was partly covered with wax to determine the removal hence the appearance of artefacts on the wafer surface.

Single etch pits were also found after etching with the OPE C. The defect density was relatively high - about 1×10^6 cm^{-3}. But etch pits were only found after long etching times (~ 20 h) and could have been caused by contaminants, particles or gas bubbles which adsorb at the silicon surface. No etch pits were found in the residual epitaxial-layer in the fragments etched with OPE F (table 20). It seems that long etching times generally lead to the formation of artefacts. The H_2O_2/HF ratio could also have an influence on the formation of these artefacts. OPE B delineated very few defects and OPE F none at all. It seems that the formation of artefacts is favoured by a high hydrogen peroxide and a high hydrofluoric acid content. Etching solutions that have a high proportion of both hydrogen peroxide and hydrofluoric acid appear to favour the formation of artefacts.

5.8 Experimental results on silicon substrates

The OPE C and D have the capability to make vacancy agglomerates (D-defects) visible. Silicon substrates from 2 different sources and types (Table 21) were etched with the OPE C and their defect densities compared to those obtained with the original (undiluted) Secco reference. 6 – 8 pieces of each type were etched with the OPE C and the Secco solution. The fragments were taken from the centre of the wafers where the incidence of voids is higher than at the edge (see chapter 3). The fragments were etched to remove a 2 -3 µm thick layer in each case. The Secco solution was not stirred during the etching process.

Material used	Thickness (µm)	Diameter (mm)	Orientation	Doping/ resistivity (Ωcm)
CZ-bulk Supplier 1	740	300	(100)	p-doped 20-25
CZ-bulk Supplier 2	740	300	(100)	p-doped 20-25
NPC (Nearly Perfect Crystals) Supplier 2	740	300	(100)	p-doped 20-25

Table 21: Physical properties of the silicon substrates used.

The fragments were then rinsed with DI-water dried in a flow of nitrogen and examined under an optical light microscope. In the Secco-etched fragments vacancy agglomerates appeared as small circular pits (figure 67). The Secco solution produces also a characteristically wedge shaped etch pattern, called flow pattern, at the crystal surface (figure 67). The formation of flow patterns is caused by evolution of hydrogen during the etching process (chapter 4.4).

At first a small hydrogen bubble located at the etch pit is formed. During the etching process the gas bubble grows and moves upwards across the silicon surface. The hydrogen ambient interferes the etching process which leads to the formation of the flow pattern [45].

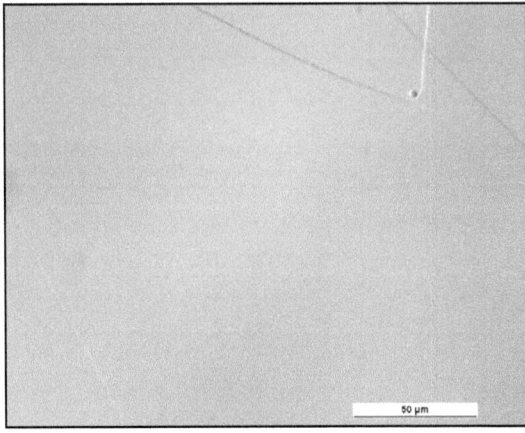

Figure 67: CZ-Bulk (NPC Fresh, Supl. 2) after etching with the Secco solution. Etching time: 16 minutes, removal: 12μm. A characteristic pattern on the wafer surface, the so called "flow-pattern", can be clearly seen.

The OPE C-etched substrates displayed voids (D-defects) delineated as square-shaped etch pits (figure 68) as mentioned before.

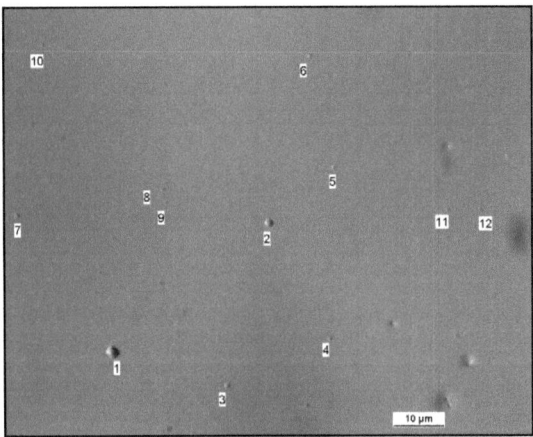

Figure 68: Square-shaped etch pits in CZ-Bulk (NPC Fresh, Supl. 2) after etching with OPE C. Etching time: 24h, removal: ~ 2μm.

The etch pit densities found after etching with OPE C are 100 to 1000 times higher than those found after etching with the Secco reference (table 22).

Material used	Etching solution used	Etching time	Removal [μm]	Defect density [cm^2]	Defect density [cm^3]	Remarks
CZ-Bulk Supl. 1	Secco	4 min.	3μm	2841	2.36 x 10^6	homogeneous removal, Flow Patterns
	OPE C	24h	2μm	572328	2.915 x 10^9	
CZ-Bulk Supl. 2	Secco	4 min.	3μm	4318	3.61 x 10^6	homogeneous removal, Flow Patterns
	OPE C	24h	2μm	130896	8.686 x 10^8	
NPC Fresh Supl. 2	Secco	4 min.	3μm	3068	2.51 x 10^6	homogeneous removal, Flow Patterns
	OPE C	24h	2μm	73 572	3.813 x 10^8	

Table 22: Defect densities found after etching with the OPE C and the Secco reference in different silicon substrates. The removal was calculated.

5.9 Experimental results on SOI material

SOI wafers consist of a thin layer of silicon built on top of a SiO$_2$ layer which acts as an insulator (figure 69). Electronic devices are built in this thin SOI layer (chapter 2).

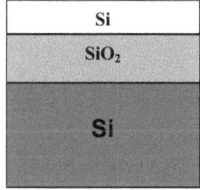

Figure 69: Structure of SOI.

The thickness of the silicon layer normally ranges from 50 to 200 nm but it is also possible to produce thin SOI with a layer thickness of less than 50 nm. The properties of the SOI layer depend on the crystal quality of the starting material. Czochralski grown silicon wafers are normally used as the starting material for the Smart-Cut^{TM} and the SIMOX processes. Such substrate wafers contain different microdefects like A/B-Swirl defects, vacancy agglomerates (D-defects) or oxygen precipitates (chapter 3). In principle these defects are transferred to the

SOI layer. During the various stages of processing additional crystal defects may also be introduced. In SIMOX material oxygen (O^+) ions are implanted into a silicon wafer at the high dose of approximately 1.8×10^{18} O^+/cm^2, using energies around 200 keV (chapter 2). The formation of SiO_2 requires a free volume which can be created by consumption of vacancies or by formation of interstitial silicon atoms (chapter 3). During the annealing steps which follow oxygen implantation, self-interstitials may agglomerate into extended defects like dislocation loops or stacking faults [11].

The density of dislocations in standard SIMOX material ranges from 1×10^4 to $5 \times 10^5/cm^2$. The stacking fault density ranges from 10^4 to $10^6/cm^2$. Tetrahedral, square-shaped or prismatic stacking faults with a size of 50 nm to 200 nm are formed [46].

In Smart-Cut™ material the crystal defects inside the thin silicon layer originate mainly from the CZ-wafer which is used to form the top layer of the SOI structure.

Such defects can be avoided by using epitaxial-wafers or by annealing at a high temperature in a hydrogen or argon atmosphere. Crystal defects may also be introduced during the thinning and polishing steps. The SOI wafers are normally heated in an oxygen atmosphere to produce a thin SiO_2 layer on top of the wafer. The silicon-dioxide is then removed by treating with HF. This procedure is used to thin the SOI layer and to polish the surface. Depending on process parameters like annealing time, oxygen concentration and/or temperature, oxidation induced stacking faults (OSF) may also be introduced into the silicon top layer.

The Organic Peracid Etches have low removal rates ranging from 0.6 nm/minute to 1.8 nm/minute. They were therefore deemed suitable for the delineation of defects in standard SOI material and thin SOI. The different OPE mixtures were tested on standard Smart-Cut™ material, on thin SOI and also on SIMOX material. The Secco diluted 1 solution was used as a reference. 6 to 8 fragments of each type of SOI with an area of approximately 1cm square were used for each OPE mixture. Only fragments from the center of the SOI wafer were normally used. After etching the wafer-fragments were rinsed with DI-water and dried in a flow of nitrogen. The SOI fragments were always etched down to a residual SOI layer thickness of approximately 30 nm. The thickness of the SOI layer was determined before and after etching with an ellipsometer. The defect densities were determined using an optical light microscope and the etched figures were counted at magnifications of 500 or 1000 times.

The diameters of the etched figures were increased by dipping the wafer pieces briefly into a concentrated hydrofluoric acid solution (figure 70). The etched figures are enlarged due to dissolution of the underlying silicon dioxide (figure 70).

Because of the short etching times a subsequent dip in HF is essential for fragments etched with the Secco diluted solution as defects would otherwise not be visible.

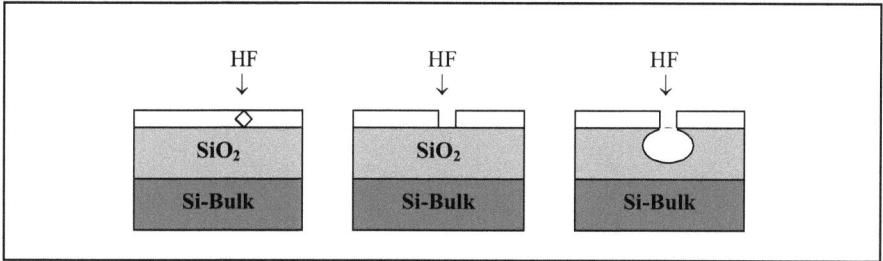

Figure 70: Effect of the dip in HF.
At first a crystal defect is attacked preferentially by the etching solution and leads to the formation of a small pit. An etch pit can be thought of as a pipe that extends from the silicon surface down to the underlying SiO_2 layer. The hydrofluoric acid flows through the pipe to the underlying silicon dioxide which is then dissolved isotropically and results in an increase in the diameter of the etch pit.

5.9.2 Influence of the hydrofluoric acid content on the size of the etched figures

Four different etching solutions with increasing HF content based on the OPE A mixture were prepared. All solutions had approximately the same peracetic acid content (~ 1.6 mol/l). The composition of the etching solutions is shown in table 22. All experiments were performed at room temperature (20°C) on standard Smart-Cut^{TM} material with an initial thickness of ~ 62 nm. An increase in HF content had no influence on the removal rate but the diameter of the etched figures increased almost linearly with increasing hydrofluoric acid content (figure 66 and table 23). In the case of the OPE A etched figures were only recognizable after an additional dip in HF. A HF content of ~ 10 vol % is sufficient to produce well-developed etched figures which are easily detected (table 23 and figure 66).

Organic Peracid Etches

Etching solution	HF content (vol %)	Etching time (minutes)	Initial thickness of the SOI layer (nm)	Removal (nm)	Diameter of the etched figures (µm)
OPE A	7.2	45	62	14 ± 0.7	0
OPE 20	11	45	62	15 ± 1.8	0.76 ± 0.03
OPE 21	14.5	45	62	15 ± 1.1	1.06 ± 0.1
OPE B	21.5	45	62	14 ± 1.8	2.52 ± 0.21

Table 23: Etching solutions with varying HF concentrations and the diameters of the corresponding etched figures.

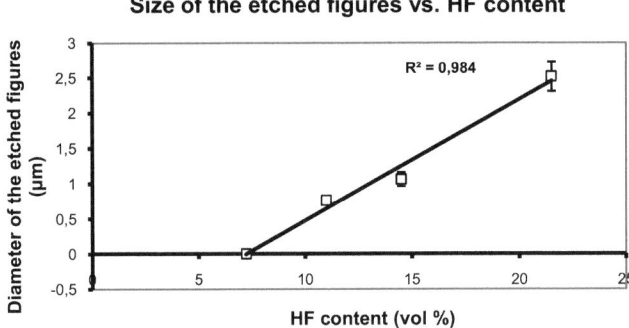

Figure 71: Diameter (r) of the etched figures as a function of the hydrofluoric acid content. The diameter increases with increasing HF content.

The size of the etched figures is also dependent on the duration of the etch. Long etching times also resulted in larger etched figures. The SOI-layer contains crystal defects of various sizes located at different depths. At first defects located near the crystal surface will be attacked. The pits formed are enlarged during the etching process (figure 72). Defects located near the buried oxide-layer will be revealed later on and appear as pits with a small diameter. The depth distribution of crystal defects in the SOI-layer will be discussed in chapter 5.13.

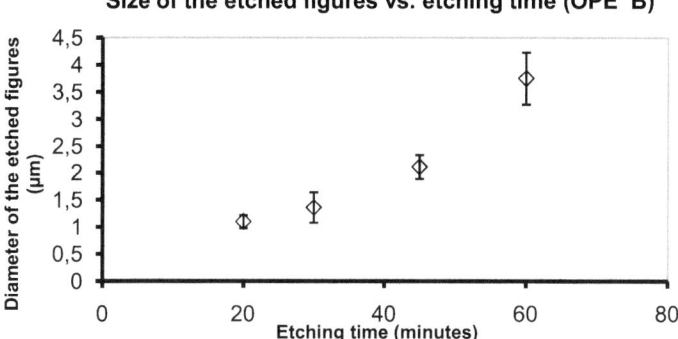

Figure 72: The diameter of the etched figures as a function of etching time shown for the OPE B.

When the OPE A and F which have a low hydrofluoric acid content are used, a subsequent dip in HF is necessary to enlarge the etched figures. This is not necessary after the OPE B, C and D etches. The long etching times in these HF-rich mixtures lead to well-developed and easily recognizable etched figures.

5.9.3 OPE tested on standard Smart-Cut^{TM} material

The OPE C and F were tested on standard Smart-Cut^{TM} material with an initial thickness of the SOI layer of ~ 145 nm, and a thickness of the underlying SiO_2 layer of about 150 nm.
A subsequent dip in HF was performed on fragments etched with the Secco diluted 1 reference and the OPE F.
All solutions were able to delineate crystalline defects in the SOI-layer. After etching defects appear as light or dark spots (figure 73-75).
In the case of the dark spots the buried oxide was dissolved completely and the etching solution had reached the substrate.

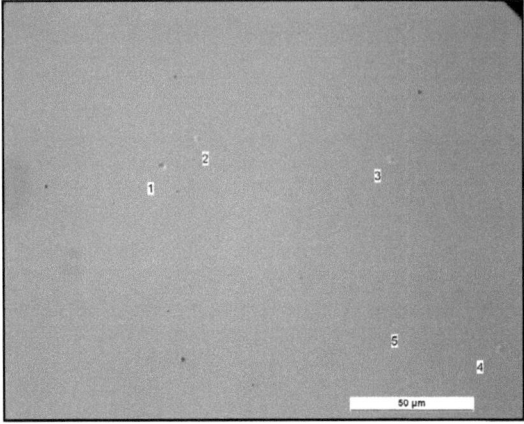

Figure 73: Standard Smart-Cut™ SOI material after etching with the Secco diluted 1 reference. Etching time: ~ 3 min., initial thickness: 145 nm, thickness after etching: ~ 30 nm, subsequent dip in HF: 60s, magnification: 500 times. Defects are delineated as small circular etched figures.

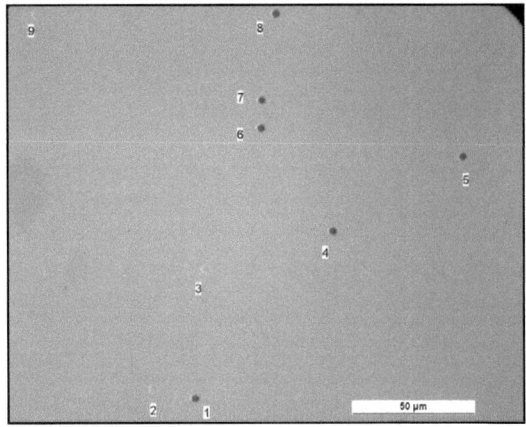

Figure 74: Standard Smart-Cut™ SOI material after etching with the OPE F. Etching time: ~ 60 minutes, initial thickness: 145 nm, thickness after etching: ~ 30 nm, subsequent dip in HF: 60s, magnification: 500 times. The etched figures look similar to those delineated by the Secco diluted reference.

Organic Peracid Etches

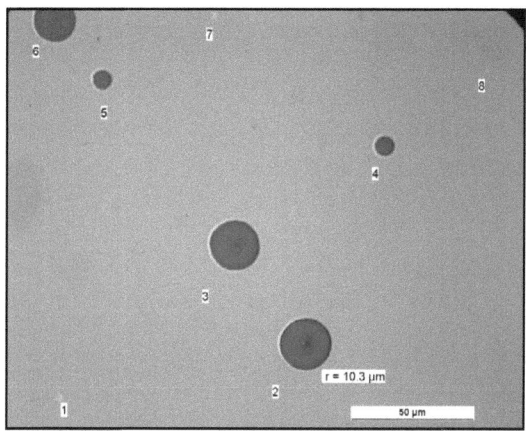

Figure 75: Standard Smart-Cut™ SOI material after etching with the OPE C. Etching time: ~ 90 min., initial thickness: 145 nm, thickness after etching: ~ 35 nm, magnification: 500 times. The etched figures are larger due to the long etching time. The depth distribution of the defects could be recognized clearly! Defects located near the surface are enlarged progressively during the etching process. Defects located near the BOX/SOI interface are attacked relatively late. The etch pits are therefore smaller.

The defect densities determined for the OPE C and F are 3 to10 times higher than those found after etching with the Secco diluted reference (table 24 and figure 76).

Etching solution used	Etching time	HF Dip	Thickness after etching (nm)	Defect density/cm^2
Secco diluted 1	**3 min.**	**60 s**	**31 ± 1.2**	**3.4 x 10^3 ± 848**
OPE C	90 min.	-	35.6 ± 4.1	4.72 x 10^4 ± 2.3 x 10^4
OPE F	60 min.	60 s	30 ± 1	1.1 x 10^4 ± 3.4 x 10^3

Table 24: Defect densities found after etching with the Secco diluted 1, OPE C and OPE F. The standard Smart-Cut™ SOI-fragments were etched down to a layer thickness of ~ 30 nm. All values are mean values. Six wafer pieces were used for each solution.

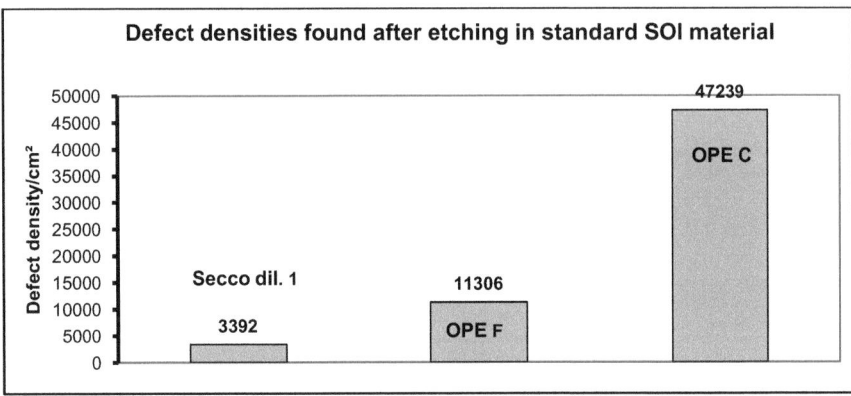

Figure 76: Comparison of the etched figure densities.
The defect densities which were obtained after etching with the OPE C and OPE F are generally higher than those found after etching with the Secco diluted reference.

The OPE C, D and F were also tested on standard Smart-Cut^{TM} SOI materials containing oxidation induced stacking faults that arise during processing. Two different materials, one with a high OSF-density and the other one with a low OSF-density, were used. All the solutions were able to produce etched figures. The OSF appear as characteristically oval shaped etch features with two individual pits in the centre. All double pits have the same length which is also characteristic of oxidation induced stacking faults (figure Figure 77-Figure 79).

Organic Peracid Etches

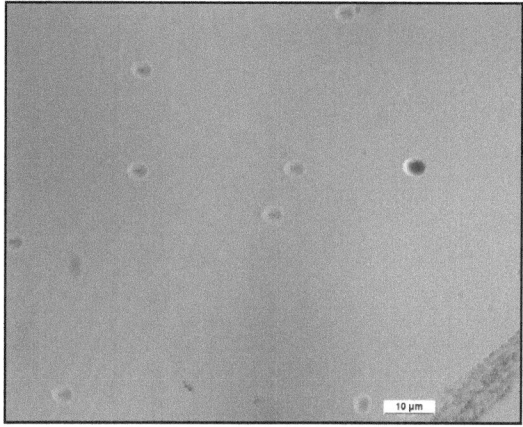

Figure 77: SOI material with a high OSF density. This fragment was etched with the Secco diluted 1. Etching time: 90 s, subsequent dip in HF: 45 s, magnification: 1000x. The OSFs are revealed as oval shaped double pits. All etched figures have roughly the same length and shape.

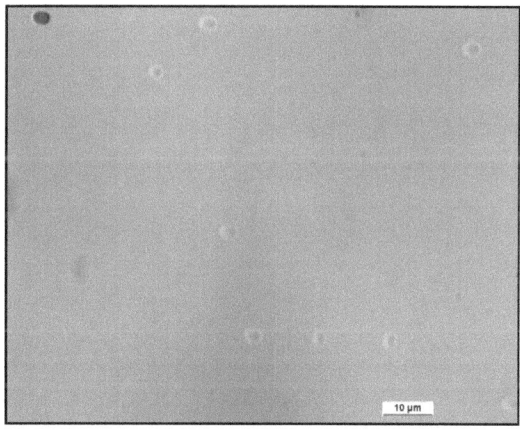

Figure 78: Oxidation-induced stacking faults in a SOI fragment etched with OPE F. Etching time: 30 min., subsequent dip in HF: 45 s, magnification: 1000 times. The double pits appear similar to those found after etching with the Secco diluted 1 reference.

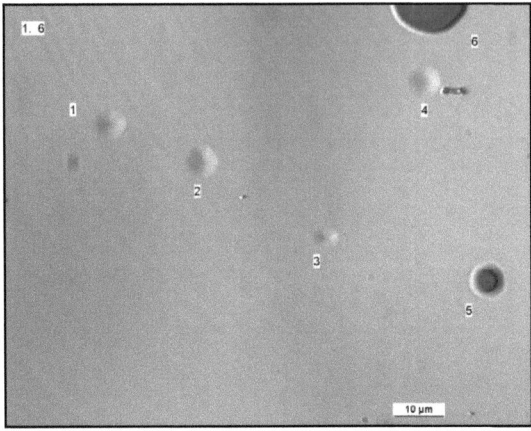

Figure 79: The OPE C is also able to reveal oxidation induced stacking-faults. The optical micrograph shows a SOI fragment etched for 50 minutes with the OPE C. A subsequent dip in HF was not necessary. Etch pits 5 and 6 have been underetched, the others, pits 1,2 and 3 are more or less round.

To compare defect densities wafer pieces were etched with the Secco diluted (reference) the OPE C, D and OPE F. Eight wafer pieces were used for each solution. In the case of the OPE F and the Secco diluted reference this was followed by a dip of 45 s in HF and the resulting double pits were oval. All etch pits had roughly the same length and shape (figure 77 and 78). The OPE C produces oval shaped double pits which are rather round (figure 79). The pits 5 and 6 apparently did not originate from the OSFs. These etch pits are probably from defects transferred from the CZ-material to the silicon top layer of the SOI-structure during the Smart-Cut^{TM} process. The etched figures that are obtained after treatment with the OPE D are different regarding their size and shape. The two dark dots in the centre cannot be detected. Therefore it is not possible to distinguish between etch pits caused by OSF and the other etch pits (figures 80 and 81).

Organic Peracid Etches

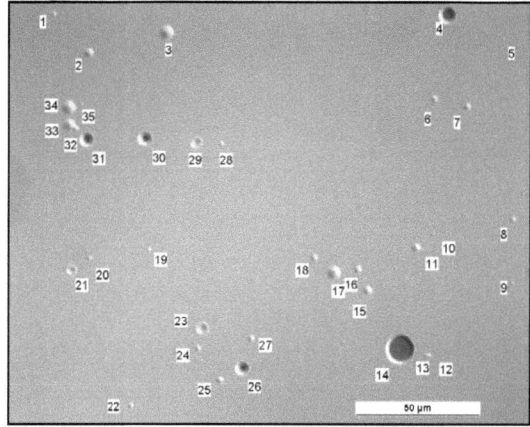

Figure 80: Different etched figures found in a SOI fragment etched with OPE D.
Etching time: 50 min., removal: 63 nm, magnification: 500 times. The pits are clearly not the characterisitic oval-shaped double pits produced when OSF sites are etched.

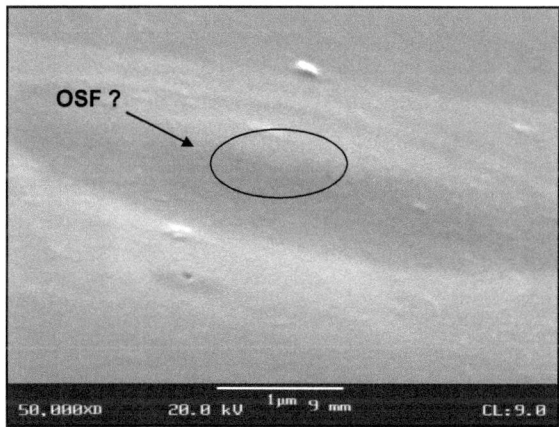

Figure 81: SEM image of an etch pit found after treatment with OPE D.
Etching time: 50 min., removal: 63 nm, magnification: 50000 times. Two individual pits in the centre cannot be seen clearly.

Etching solution used	Etching time	Dip in HF	Thickness after etching (nm)	OSF density/cm^2	Total defect density/cm^2
Secco diluted 1	90 s	45 s	30 ± 3.7	8.5 x 10^4 ± 7.5 x 10^3	8.5 x 10^4 ± 7.5 x 10^3
OPE C	50 min.	-	30 ± 1.7	7.7 x 10^4 ± 1.9 x 10^4	1 x 10^5 ± 2 x 10^4
OPE D	50 min.	-	30 ± 3.9	-	1.1 x 10^5 ± 2.5 x 10^4
OPE F	**30 min.**	**45 s**	**28 ± 1.45**	**8.2 x 10^4 ± 4 x 10^3**	**8.2 x 10^4 ± 4 x 10^3**

Table 25: Defect densities obtained on three different OPE mixtures tested on standard Smart-Cut^{TM} material with a high OSF density.

The Secco diluted 1 solution was used as a reference. All values which are shown above are mean values. The OSF densities found after etching are all within a narrow range. The OPE C and D are also able to reveal crystal defects which are not caused by oxidation induced stacking faults. Therefore the total defect density is higher than that obtained with the Secco diluted 1 reference and the OPE F.

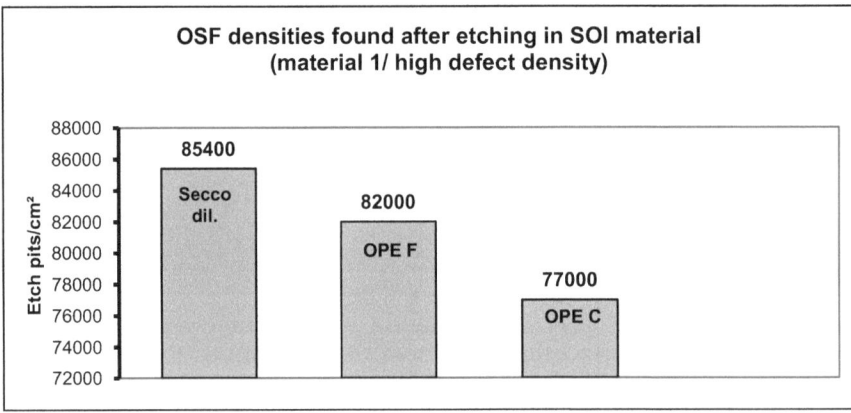

Figure 82: OSF densities found after etching with the Secco diluted reference and the OPE C and F.

Organic Peracid Etches

Etching solution used	Etching time	Dip in HF	Thickness after etching (nm)	OSF density/cm^2	Total defect density/cm^2
Secco diluted 1	90 s	45 s	35 ± 1.2	2 x 10^3 ± 0.57 x 10^3	2 x 10^3 ± 0.57 x 10^3
OPE C	50 min.	-	37 ± 2	4 x 10^3 ± 3.9 x 10^3	4.8 x 10^3 ± 2.7 x 10^3
OPE D	50 min.	-	35 ± 2.4	-	1.7x 10^4 ± 1.4 x 10^4
OPE F	30 min.	45 s	27 ± 1	2.3 x 10^3 ± 1 x 10^3	2.3 x 10^3 ± 1 x 10^3

Table 26: The OPE mixtures C, D and F were also tested on standard Smart-Cut^{TM} material with a low OSF density. The OSF density found after etching with the OPE C is 2 times higher than those found after etching with the Secco diluted reference and the OPE F.

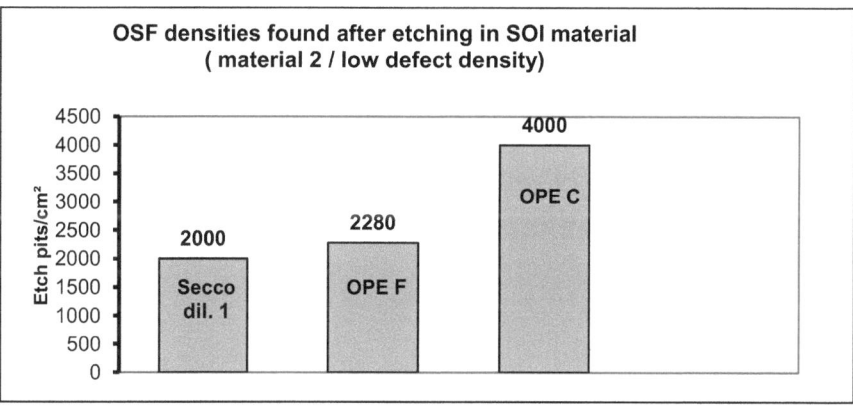

Figure 83: Defect densities found after etching with the Secco diluted reference and the various OPE mixtures.

OPE C and D apparently also reveal crystal defects other than OSFs. This is why the total defect densities are higher than the OSF densities (table 25 and 26). The etched figures obtained after treatment with the OPE D are rather round and are also of different sizes. Two individual pits in the centre of an etched figure are not clearly recognizable. It is not possible to distinguish between etch pits caused by OSF which have all the same size and etch pits caused by defects transferred from the starting material. The OSF densities are similar to the total defect densities obtained with the Secco diluted 1 reference and the OPE F (table 25 and 26). In SOI material with high OSF densities, the OSF densities determined for the OPE C, F and the Secco diluted 1 solution were all within a narrow range, those obtained with the OPE F and the Secco dilute reference being nearly the same (Fig. 82).

On the other hand, in SOI material with relatively low OSF densities the OSF densities found after etching with the OPE C were approximately two times higher than those obtained with the Secco diluted 1 reference mixture (Fig. 83). Again the OSF density obtained with OPE F is very close to that of the Secco dilute reference.

5.9.4 Experimental results on thin SOI

The same three Organic Peracid Etches and the OPE A were used to etch thin Smart-Cut^{TM} material with an initial layer thickness of 17 - 21 nm. For each mixture six different fragments were used, taken from the edge and the center of the wafer (figure 84).

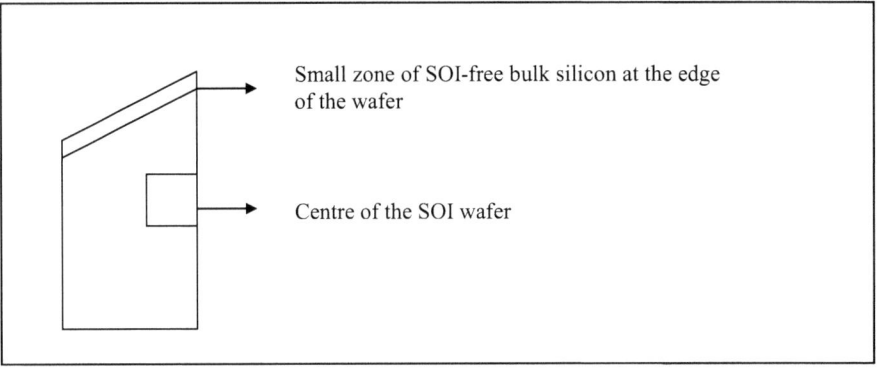

Figure 84: The samples were taken from the centre and the edge of the wafer.

The samples taken from the centre of the wafer were etched with the OPE A, C, D, F and the Secco diluted 1 which was used as the reference. The initial and residual thicknesses of the SOI layer were determined for each fragment with an ellipsometer. The etching times were chosen to remove half of the initial SOI layer. In case of the OPE A, F and the Secco diluted 1 reference an additional dip in HF of 30 s was always performed after etching. Defects were revealed as small circular etch pits and in the case of the OPE A also as square-shaped pits (figure 85 – 87).

Organic Peracid Etches

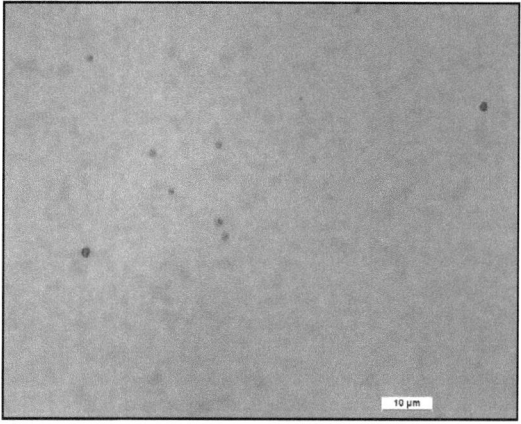

Figure 85: Thin SOI after etching with the Secco diluted 1 solution.
Etching time: 10 s, initial thickness: ~ 17 nm, thickness after etching: ~ 8.5 nm,
magnification: 1000 times. Defects are revealed as small circular pits.

Figure 86: Thin SOI after etching with the OPE D.
Etching time: 5 min., initial thickness: ~ 20 nm, thickness after etching: ~ 12 nm,
magnification: 1000 times. Defects are delineated as circular pits.

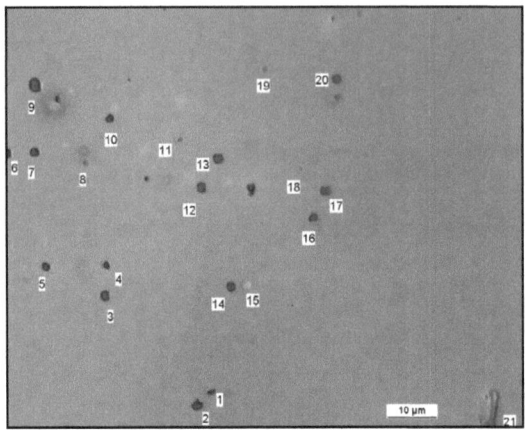

Figure 87: Crystalline defects in thin SOI material delineated with OPE A. Etching time: 5 min., initial thickness of the SOI layer: ~ 18 nm, thickness after etching: ~ 11 nm, magnification: 1000 times. Defects are delineated as circular and square-shaped etch pits.

The defect densities found after etching are all comparable except for that found after etching with the OPE A which was significantly higher (table 27, figure 88). The etch pit density determined for the OPE C agrees very well with that of the Secco diluted reference.

Etching solution used	Etching time	HF Dip	Thickness after etching (nm)	Defect density/cm^2
Secco diluted 1	10s	30s	8.4 ± 1	**9.3 x 10^3 ± 6.8 x 10^3**
OPE A	10 min.	30s	11 ± 0.5	3.64 x 10^4 ± 7.9 x 10^3
OPE C	**6 min.**	-	**11.3 ± 0.7**	**9.5 x 10^3 ± 2.1 x 10^3**
OPE D	5 min.	-	11.3 ± 0.7	5.8 x 10^3 ± 4.2 x 10^3
OPE F	**5 min.**	30s	**9.4 ± 0.3**	**1.5 x 10^4 ± 4 x 10^3**

Table 27: Defect densities in thin SOI found after etching with the different OPE mixtures and the Secco diluted 1 reference. All values are mean values.

Organic Peracid Etches

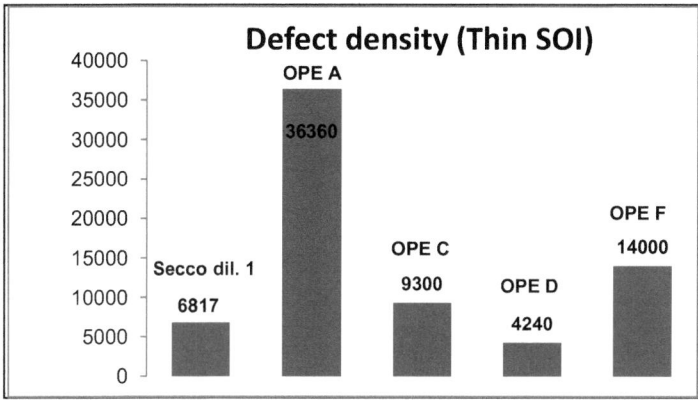

Figure 88: The defect densities obtained after etching with different formulations. Defect densities found after etching with the OPE A are approximately ten times higher than those found with Secco diluted 1.

Fragments taken from the edge of the wafer were also used. These fragments include a small zone of silicon bulk where the SOI film is absent (figure 84). The fragments were etched with the OPE A, D and the Secco diluted 1 reference. Small loops or clusters of loops were found in the bulk area after etching (figure 89 – 90). These loops are probably caused by Swirl-defects (chapter 3). Single loops or complicated arrangements of loops are obtained.

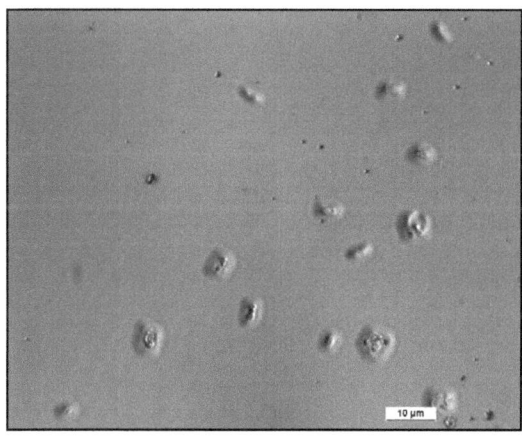

Figure 89: Swirl defects found after etching with the Secco diluted 1. Etching time: 10s, removal ~ 10 nm, magnification: 1000 times.

Organic Peracid Etches

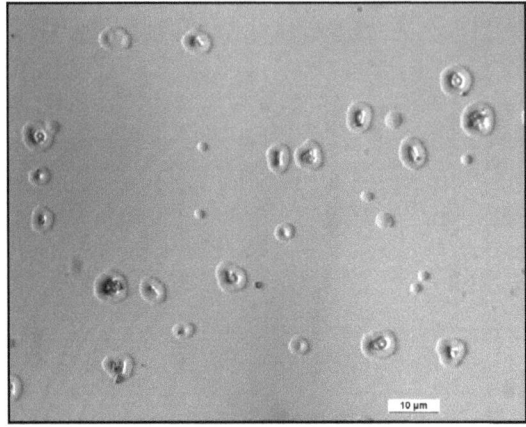

Figure 90: Swirl defects revealed with the OPE A.
Single loops or clusters of individual loops are obtained. The Swirl-defects were delineated in a small zone of silicon bulk at the edge of the SOI wafer. Etching time: 15 min., removal ~ 7 nm, magnification: 1000 times.

The small round etch features could probably have been caused by B-defects which can be thought of as forerunners of the Swirl-defects (chapter 3.4)
The defect densities found after etching with the OPE A and D are approximately ten times higher than those found with the Secco diluted 1 solution (table 28 and figure 91). The OPE mixtures are also able to delineate even small A-defects.

Etching solution used	Etching time	Dip in HF	Removal (nm)	Defect density/cm^2	Defect density/cm^3
Secco diluted 1	10 s	-	10 ± 1	$4.8 \times 10^4 \pm 1 \times 10^4$	$2.1 \times 10^{11} \pm 4.5 \times 10^{10}$
OPE A	15 min.	-	7 ± 1.2	$3 \times 10^5 \pm 3.4 \times 10^4$	$\mathbf{3.2 \times 10^{11} \pm 3.2 \times 10^{11}}$
OPE D	6 min.	-	10 ± 1.3	$3.25 \times 10^5 \pm 2.7 \times 10^4$	$\mathbf{3.7 \times 10^{11} \pm 3.3 \times 10^{11}}$

Table 28: Comparison of the Swirl-defect densities found after etching with OPE A, D and the Secco diluted reference.
The Swirl-defects appear only in a small zone at the edge of the wafer.

Organic Peracid Etches

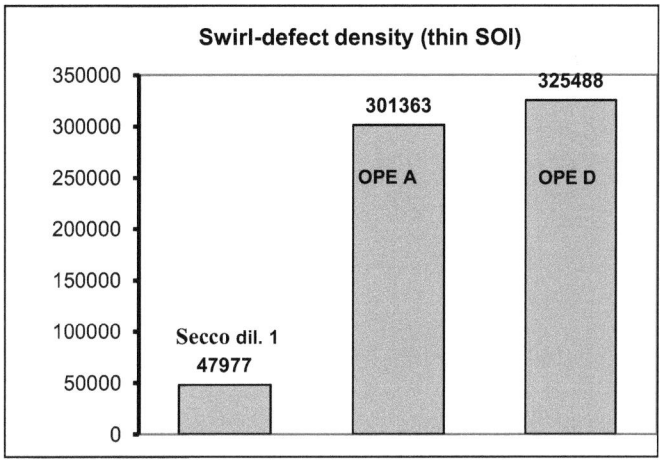

Figure 91: Swirl-defect densities obtained after etching with the OPE A and OPE D. They are approximately ten times higher than those obtained with the Secco diluted reference.

5.9.5 Experimental results on Simox material

SOI-material produced by the Separation-by-Implanted-Oxygen technique (SIMOX) differs clearly from the standard Smart-Cut^{TM} material. The implantation of oxygen ions produces additional crystal defects in the SOI layer. Experiments were also conducted using Organic Peracid Etches to delineate crystal defects in SOI material obtained by the SIMOX-process. SIMOX material with an initial SOI thickness of ~ 80 nm was used.

The mixtures used were OPE C, D, F and the Secco diluted 1 which served again as a reference. Each mixture was tested on a separate batch of 8 wafers. Etching with Secco diluted 1 and OPE F were followed by a dip in HF.

All solutions were found to delineate defects in the SOI layer (figure 92 – 94).

Crystal defects are revealed as circular etch pits (figure 92 – 94). The OPE C and D produce larger etch pits due to the long etching times (figure 94).

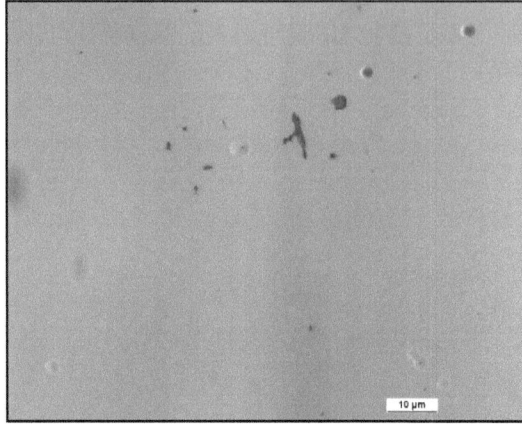

Figure 92: SIMOX SOI material after etching with the Secco diluted 1.
Etching time: 1 min., additional dip in HF: 60 s, initial thickness of the SOI-layer: ~ 80 nm,
thickness after etching: ~ 27 nm, magnification: 1000 times.

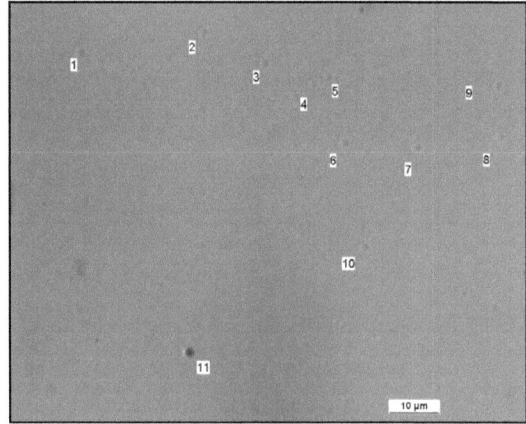

Figure 93 : Crystal defects in SIMOX SOI found after etching with OPE F.
Etching time: 26 min., additional dip in HF: 60 s, initial thickness of the SOI layer: ~ 80 nm,
thickness after etching: ~ 25 nm, magnification: 1000 times.

Organic Peracid Etches

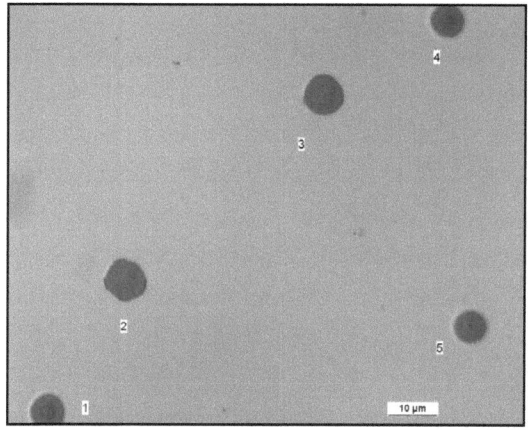

Figure 94: Crystal defects in SIMOX SOI found after etching with OPE C. Etching time: 32 min., initial thickness of the SOI layer: ~ 80 nm, thickness after etching: ~ 29 nm, magnification: 1000 times. The etched figures are enlarged due to the long etching times and the high HF content of the solution.

The defect densities found after etching with the Organic Peracid Etches C, D and F are approximately ten times higher than those etched with the Secco diluted reference (table 29 and figure 95).

Etching solution used	Etching time	Dip in HF	Thickness after etching (nm)	Defect density/cm^2
Secco diluted 1	**60 s**	**60 s**	**27 ± 0.6**	**4.52 x 10^3 ± 1.13 x 10^3**
OPE C	32 min.	-	30 ± 1	3.28 x 10^4 ± 1.3 x 10^4
OPE D	25 min.	-	31 ± 0	3.49 x 10^4 ± 1.47 x 10^4
OPE F	25 min.	60 s	28 ± 1.45	5.45 x 10^4 ± 1.48 x 10^3

Table 29 Table 28: Comparison of defect densities found after etching in SOI material produced by the SIMOX technique. All values are mean values.

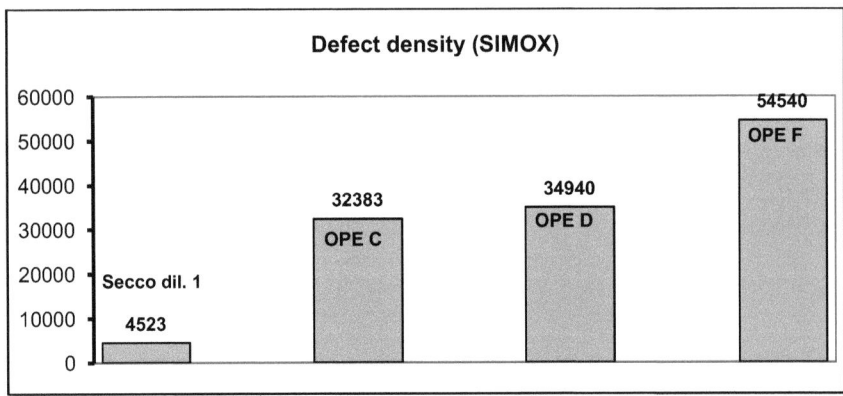

Figure 95 : Defect densities in SIMOX material found after etching with the OPE mixtures C, D and F.

5.9.6 Depth distribution of crystal defects in the SOI layer

The effect on defect density of the amount removed with etching was also studied. Standard Smart-Cut^{TM} SOI material with an initial thickness of ~ 88 nm containing "as grown defects" which were transferred from the CZ-Wafer to the SOI layer was used for this experiment. The SOI fragments were etched down with the OPE D to different thicknesses. Because of its low removal rate it is possible to etch down the SOI layer step by step. The defect densities were determined after each etching step. The crystal defects are different in regard to their sizes and depth distribution (figure 96). The defect density per cm^2 should increase on removal (chapter 5.9.2) .

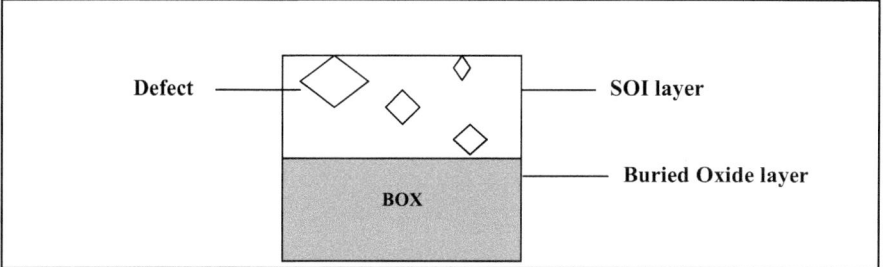

Figure 96: Depth distribution of crystal defects in the SOI layer.
The size of the defects also varies.

The experimental results are shown on table 30. After seven minutes of etching only isolated and very small etched figures were obtained (figure 97) which means, that these defects do not extend as far as the underlying buried oxide. The size of the defects in the vertical direction should be in the range of 50 – 60 nm. Well developed etched figures are obtained after 25 minutes etching time. With longer etching times the size of the etched figures increases due to the underetching of the buried oxide and even small defects are clearly visible (figure 98 and 99).

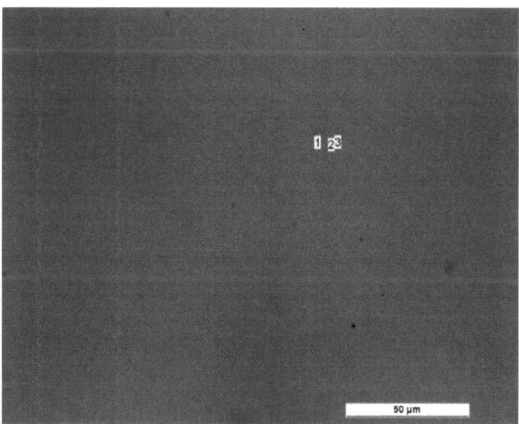

Figure 97 : Small etched figures found after etching with the OPE D.
Material used: Standard Smart-Cut^{TM} SOI, etching time: 7 min., initial thickness: ~ 88 nm, thickness after etching: ~ 77 nm, magnification: 500 times.

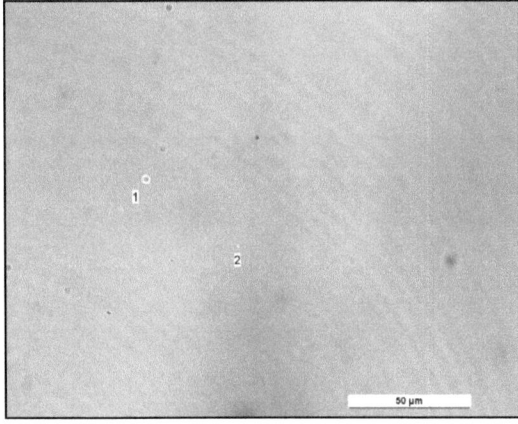

Figure 98: Standard Smart-Cut™ SOI, etched with OPE D for 25 min., initial thickness: ~ 88 nm, thickness after etching: ~ 42 nm, magnification: 1000 times.

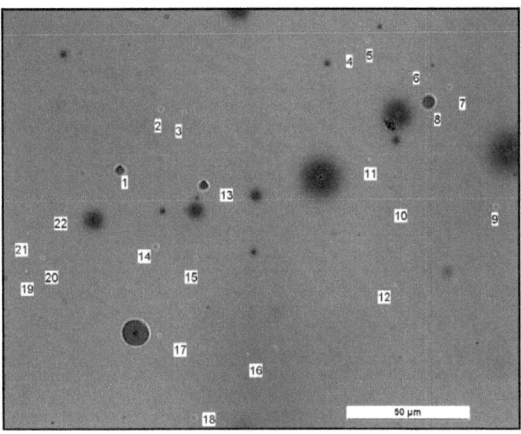

Figure 99: The defect density increases with increasing removal. Even small defects are revealed by the preferential etch. Material used: Standard Smart-Cut™ SOI, etching solution used: OPE D, etching time: 35 min., initial thickness: ~ 88 nm, thickness after etching: ~ 30 nm, magnification: 500 times.

The crystal defects in the SOI layer were always delineated as circular etched figures. As expected the defect density per cm^2 increased with increasing removal (table 30).

Etching time (min.)	Thickness after etching (nm)	Removal (%)	Defect density/cm^2	Defect density/cm^3
7	76.5 ± 0.5	13	1460 ± 854	1.3 x 10^8 ± 7.4 x 10^7
15	62.4 ± 1	30	8621 ± 6863	3.4 x 10^8 ± 2.8 x 10^8
25	45.5 ± 2.3	49	9658 ± 3928	2.3 x 10^8 ± 9 x 10^7
35	32 ± 3.5	64	29161 ± 13417	5.2 x 10^8 ± 2.4 x 10^8

Table 30: Defect densities obtained after different etching times. All values are mean values.

Figure 100 shows the dependence of the defect density on removal. After more than half of the SOI layer was removed a sharp increase in the defect density was observed. Most of the crystal defects which are located at different depths of the SOI layer have now reached the buried oxide layer.

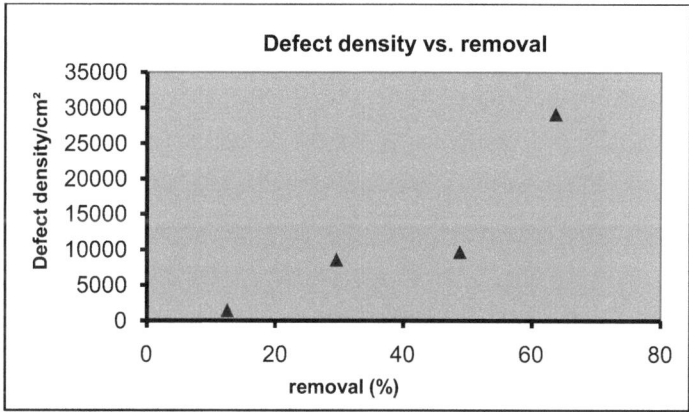

Figure 100: Defect density as a function of removal.

5.10 Detailed characterisation of the etched figures by atomic force microscopy (AFM), scanning electron microscopy (SEM) and transmission electron microscopy (TEM)

It is known that the Secco diluted solutions always produce etch pits in the SOI layer. The diameter of these etch pits is only dependent on the duration of the dip in HF that follows. Different methods like AFM, SEM or TEM were used to investigate the nature of the etched figures obtained after treatment with the Organic Peracid Etches.

5.10.1 AFM investigations

Different SOI materials were etched with the OPE mixtures and the etched figures were then characterized in detail using atomic force microscopy (AFM) in the tapping-mode. At first the OPE A was tested on standard Smart-Cut^{TM} SOI-material with an initial thickness of the SOI layer of ~ 62 nm and a buried oxide layer of ~ 150 nm. The SOI layer was etched down to a thickness of around 40 nm. A subsequent dip in HF was not performed. Defects revealed were round (figure 101). AFM investigations showed that the etched figures were hillocks (figure 102).

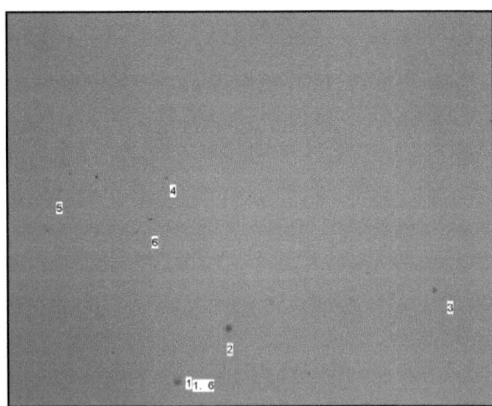

Figure 101: Standard Smart-Cut^{TM} SOI material after etching with the OPE A. Etching time: 45 min., initial thickness: ~ 62 nm, thickness after etching: ~ 40 nm, magnification: 500 times. A subsequent dip in HF was not performed! Etched figures can be recognized clearly.

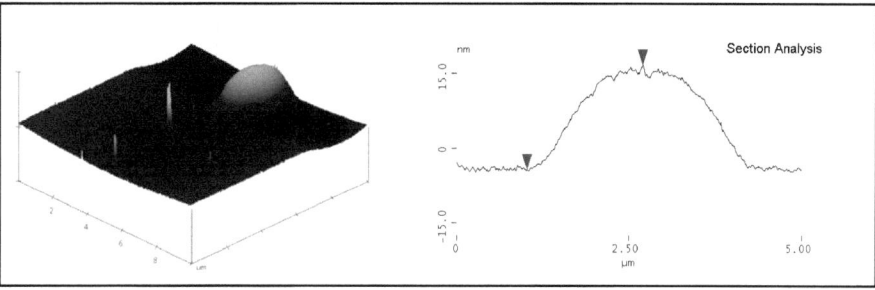

Figure 102: AFM image of an etch hillock (scan area 10 μm x 10 μm) found in standard Smart-Cut™ SOI-material after etching with the OPE A.
Initial thickness of the SOI-layer: ~ 62 nm, removal: ~ 20 nm.

Samples of the same material were also etched with the OPE B. The etched figures obtained differed clearly from those found after etching with the OPE A (figure 103). Their varying size indicates that they were located at different depth in the SOI film. They were also round but had a larger diameter, due to the higher hydrofluoric acid content of the OPE B. In contrast to the OPE A a little dot was present in the centre of each etched figure (figure 103). AFM investigations revealed that the OPE B also produces etch hillocks (figure 104).

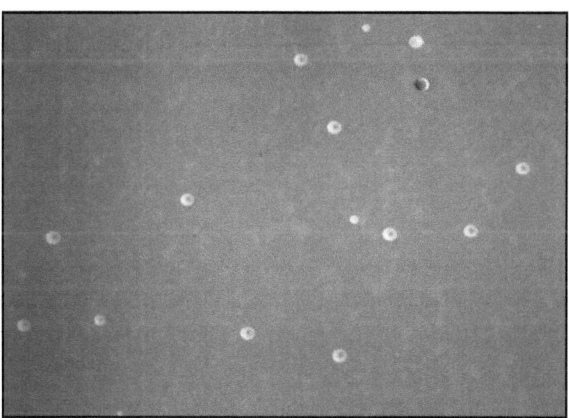

Figure 103: Standard Smart-Cut™ SOI material after etching with the OPE B.
Etching time: 45 min., initial thickness: ~ 62 nm, thickness after etching: ~ 45 nm, magnification: 500 times. Etched figures can be recognized clearly.

Organic Peracid Etches

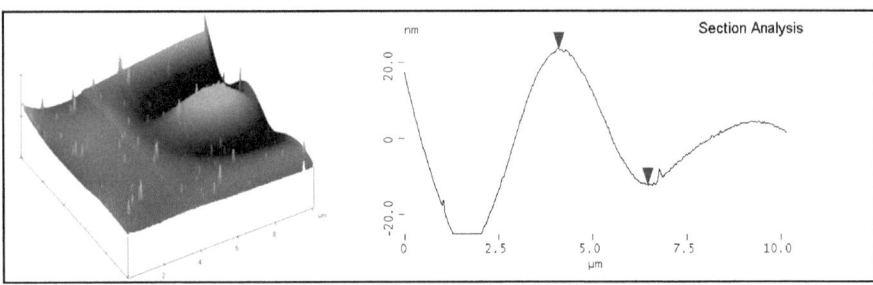

Figure 104: AFM image of an etch hillock (scan area 10 μm x 10 μm) found in standard Smart-Cut[TM] SOI material after etching with the OPE B. Initial thickness of the SOI layer: ~62 nm, removal: ~ 17 nm.

The same standard Smart-Cut[TM] SOI material was also etched with the OPE D. The OPE D contains perpropanoic acid instead of peracetic acid (table 15) and the removal rate is three times higher than that of the OPE A and B (table 18). The SOI layer was also etched down to a thickness of ~ 40 nm. The etched figures are similar to those found after etching with the OPE B (figure 105).

Figure 105: Standard Smart-CutTM SOI material after etching with the OPE D. Etching time: 15 min., initial thickness: ~ 62 nm, thickness after etching: ~ 38 nm, magnification: 500 times. The etched figures show some size variation indicating again that they were located at different depth in the SOI film.

The subsequent AFM measurements showed that the etched figures were pits in contrast to the hillocks produced by OPE A and B. Figure 106 shows the 3D image and the line scan of an etch pit found after treatment with OPE D. Depth of the etch pit: ~ 300 nm. This means that the underlying buried oxide layer was completely dissolved and the silicon substrate had also already been attacked by the etching solution. This is concluded from the depth of the etch pit exceeding the sum of the thickness of the SOI and BOX layer and the shape of the pit cross section revealed in the AFM line scan.

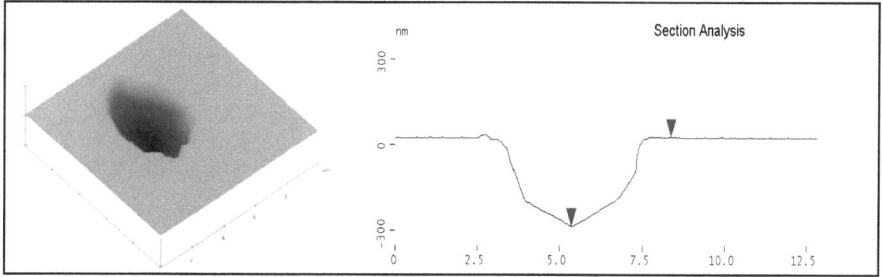

Figure 106: AFM 3D image and line scan of an etch pit (scan area 10 μm x 10 μm) found in standard Smart-Cut™ SOI material after etching with the OPE D.
Initial thickness of the SOI layer: ~ 62 nm, thickness after etching: ~ 38 nm, depth of the etch pit: ~ 300 nm.

The OPE D was also tested on Standard Smart-Cut™ SOI material with an initial SOI thickness of ~ 90 nm and on thick SOI with an initial SOI thickness of ~ 200 nm. The thick SOI was also produced by the Smart-Cut™ technique. The physical properties of the different materials used are summarized in table 31.

Material used	Initial thickness of the SOI layer [nm]	Thickness of the BOX [nm]
Standard Smart-Cut™ SOI	89	150
Thick SOI	200	400
Thick SOI (high Ω)	200	800

Table 31: Properties of the different SOI materials.

Table 32 shows the experimental results. All three SOI materials displayed etch hillocks. Very well developed etched figures were obtained on the thick SOI material (figure 107) with larger diameters due to the long etching time and the high HF content of the solution used.

Material used	Etching solution used	Etching time	Removal	Defect density/cm^2	Kind of defects
Standard Smart-Cut^{TM} SOI	OPE D	25 minutes	~ 43 nm	4240	hillocks
Thick SOI (high Ω)	OPE D	90 minutes	~ 150 nm	1700	hillocks
Thick SOI	OPE D	30 minutes	~ 50 nm	1060	hillocks

Table 32: Summary of the experimental results of Smart-Cut^{TM} SOI materials etched with the OPE D. Etch hillocks were found in all cases.

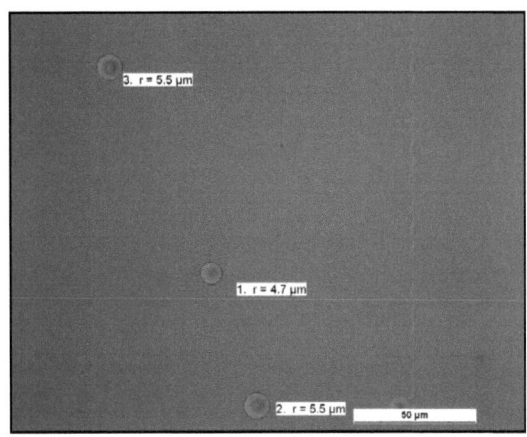

Figure 107: Etched figures in thick SOI material found after etching with the OPE D. Etching time: 90 min., initial thickness: ~ 200 nm, thickness after etching: ~ 50 nm, magnification: 500 times

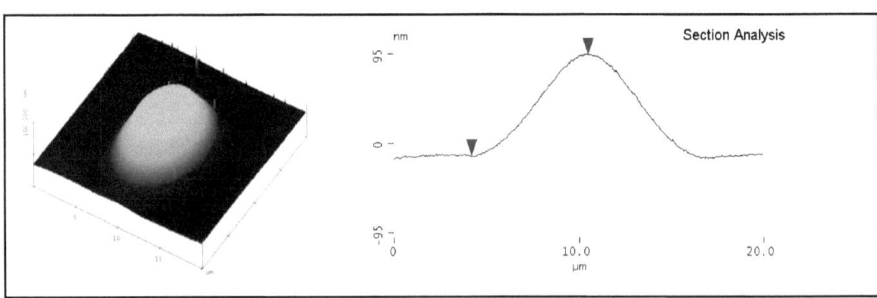

Figure 108: AFM 3D image and line scan of an etch hillock (scan area 20 µm x 20 µm) found in thick Smart-Cut^{TM} SOI material (high Ω) after etching with the OPE D. Initial thickness of the SOI layer: ~ 200 nm, thickness after etching: ~ 50 nm.

The experimental results are summarized in table 33. The Organic Peracid Etches produce both etch hillocks and etch pits depending on the experimental conditions. Long etching times, a high hydrofluoric acid content and a thin SOI layer all favour the formation of etch pits. A dip in HF after etching also leads to the formation of etch pits. Etch hillocks are more likely to be formed with short etching times and a thick SOI layer.

Material used	Etching solution used	Etching time	Initial thickness of the SOI-layer	Thickness after etching	Kind of defects
Standard Smart-Cut^{TM} SOI	OPE A	45 minutes	~ 62 nm	~ 40 nm	hillocks
Standard Smart-Cut^{TM} SOI	OPE B	45 minutes	~ 62 nm	~ 45 nm	hillocks
Standard Smart-Cut^{TM} SOI	OPE D	15 minutes	~ 62 nm	~ 38 nm	pits
Standard Smart-Cut^{TM} SOI	OPE D	25 minutes	~ 89 nm	~ 56 nm	hillocks
Thick SOI (high Ω)	OPE D	90 minutes	~ 200 nm	~ 50 nm	hillocks
Thick SOI	OPE D	30 minutes	~ 200 nm	~ 150 nm	hillocks

Table 33: Summary of the experimental results. The Organic Peracid Etches produce both pits and hillocks.

5.10.2 Characterisation of the etch hillocks by SEM and TEM

The formation of etch hillocks in the SOI-layer is rather unusual. Etch hillocks are normally caused by metal precipitates or agglomerates of self-interstitials (chapter 2). Detailed SEM and TEM investigations were performed at the Siemens Research-Laboratories in Munich to study the nature of these hillocks.

The thick SOI-material with an initial layer thickness of ~ 200 nm and a thickness of the buried oxide layer of ~ 800 nm (table 31) was used for these experiments.

A little hole with a diameter of ~ 250 nm was always found in the centre of the etch hillocks (figure 109).

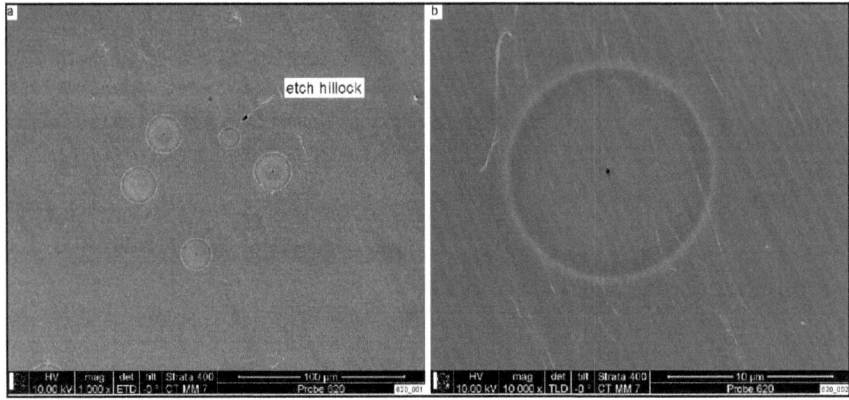

Figure 109: SEM images of etch hillocks found in thick SOI after etching with the OPE D. Etching time: 90 min., initial thickness of the SOI layer: ~ 200 nm, thickness after etching: ~ 50 nm, thickness of the BOX-layer: 800 nm. Note the little hole in the centre of the hillock in the image on the right.

The sample shown in figure 110 was then prepared for further TEM-investigations. An etch hillock was selected and a thin Pt-layer deposited by sputtering on the surface. The etch hillock was then cut through using the Dual-Beam Focussed Ion Beam (FIB) technique (figure 110). It now became apparent that the BOX below the thin SOI-layer had been removed during the OPE D etching process (figure 110).

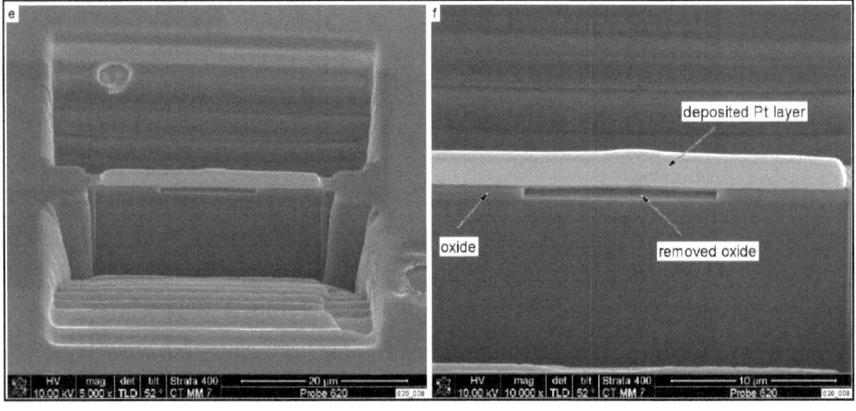

Figure 110: SEM image of a vertical cross-section of a selected etch hillock made with the FIB-technique. A cavity was found below the hillock.

The TEM images in figure 111 a-c show sputter residues from the FIB thinning process on the side-walls of the opening in the oxide. The opening in the silicon layer which was then filled with Pt used for the deposition to protect the etch hillock can be seen (figure 111 c).

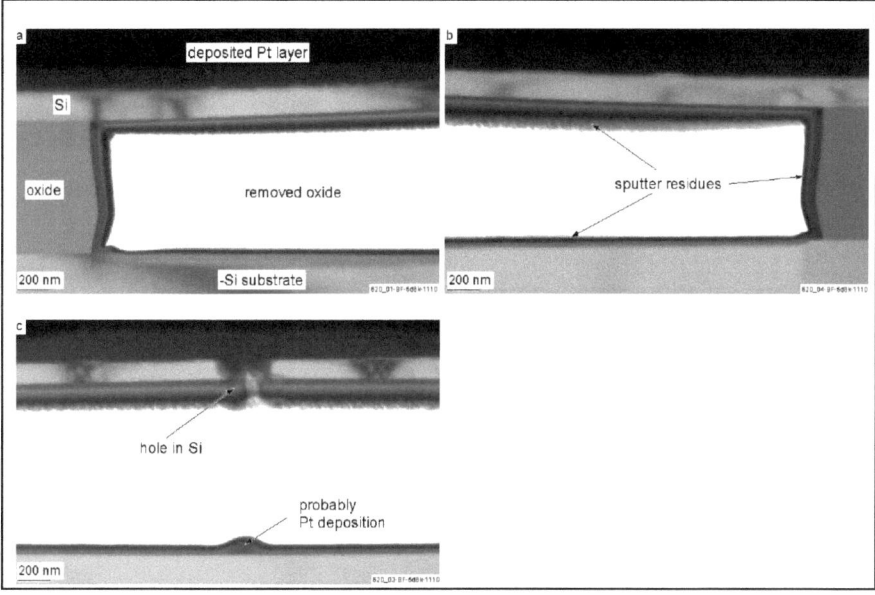

Figure 111 a-c: 200 kV X-TEM bright field image of an etch hillocks in SOI material. The underlying buried oxide layer is removed (a, b). The opening in the silicon layer which was then filled with Pt used for the deposition to protect the etch hillock can be seen (c).

A possible explanation for the formation of the etch hillocks could be the following: at first the etching solution preferentially attacks a crystal defect in the SOI-layer (figure 112). A channel leading in the SOI layer down to the buried oxide is formed (figure 112). Through this channel the etching solution reaches the buried oxide.
Because of the high HF-content of the OPE mixtures and the long etching times, the dissolution of the underlying silicon dioxide that follows is faster than the removal of the silicon layer around the defect. This effect leads to the formation of a cavity below the hillock (figure 113). Finally, the thin silicon layer above the cavity curves upwards due to some intrinsic strain originating from the SOI fabrication process (figure 113). The little hole in the centre of each etch hillock was produced by the "original" crystal defect. Its diameter of

around 250 nm corresponds roughly to that of voids (D-defects,COPs), formed during CZ crystal growth which have been transferred from the original CZ-wafer to the SOI-structure.

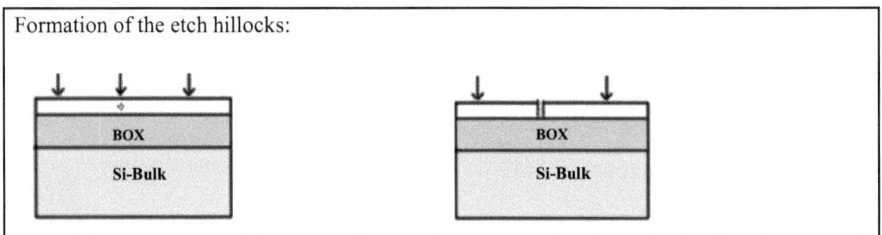

Figure 112: Preferential attack at the crystal defect and channel formation. in the SOI film.

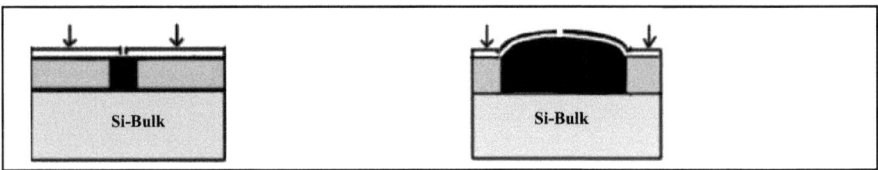

Figure 113: Dissolution of the underlying buried oxide and hillock formation.

5.11 Experimental results on sSOI material

The Organic Peracid Etches were also tested on Strained Silicon on Insulator (sSOI) material which consists of a thin strained silicon layer on top of an insulator like silicon dioxide (figure 114). Integrated circuits are built in this thin sSOI layer.
The physical properties of the sSOI material were discussed in chapter 2.6.

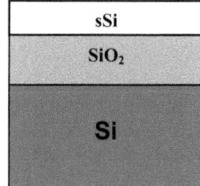

Figure 114: Structure of sSOI

The strained silicon layer contains a lot of crystal defects which include stacking faults, micro twins, pile ups and threading dislocations (TD) that arise during processing.

A strained silicon layer is obtained by epitaxial silicon deposition on a relaxed SiGe buffer layer which is referred to as a virtual substrate (chapter 2.6). The virtual substrate is formed by depositing a graded SiGe buffer on the surface of a polished silicon wafer.
The different lattice constants cause a strain inside the SiGe layer. The lattice constant of the SiGe mixed crystal will be reduced parallel to the silicon surface and increases vertically [47]. When the thickness of the SiGe layer exceeds the critical thickness h_c the strain is relieved by the formation of dislocations [48]. These dislocations are called threading dislocations and can be thought of as a superposition of an edge and a screw dislocation. An even increase of the Ge content in the graded buffer allows a minimization of threading dislocations by the reuse of existing threads [49].
The formation of further dislocations is suppressed. The final result is a relaxed SiGe layer with a relatively low ($\sim 10^5/cm^2$) TD density (figure 115). Threading dislocations cannot end inside the SiGe alloy. They pass through the crystal up to surface.

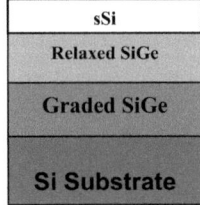

Figure 115 Strained silicon on a virtual substrate

Finally, the silicon is deposited on the virtual substrate by chemical vapour deposition. Below the critical thickness of ~ 16 nm or at elevated temperatures around 650°C, the strained silicon layer relaxes by the glide of existing threading dislocations or by the nucleation of two new threads (figure 116) [48].

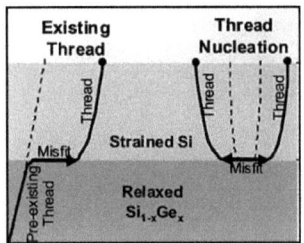

Figure 116: Relaxing of the strained silicon layer
via misfit formation by the glide of an existing TD or
the nucleation of new threading segments. Reference: *Taraschi et al Journal of the Electrochemical Society Vol. 151 (1), 2004*

The Secco diluted solutions are normally used to reveal the different crystal defects in the sSi layer. Taraschi et al. found that H_2O_2/HF/HAc mixtures which are normally used to remove the Si/Ge transfer layer in sSOI fabrication (chapter 2.6) are also able to produce etched figures in the sSi film. Defect densities of ~ 10^5/cm² were always found. It was assumed that these etched figures are caused by threading dislocations [48] in which case the Organic Peracid Etches used in this study should also be able to reveal TD in the strained silicon layer. At first the OPE A was tested on standard sSOI material with an initial thickness of the strained silicon layer of ~ 30 nm. The removal rate on sSOI is similar to that determined on SOI material. All experiments were performed at room temperature (24°C). The sSOI layer was etched down to a residual thickness of ~ 15 nm. followed by a dip in HF. Threading dislocations were revealed in the strained silicon film (figure 117 and 118).

Other crystalline defects like stacking faults or twins which also appear in the strained silicon film were not delineated. These experimental results are in accordance with those obtained by Taraschi et al.

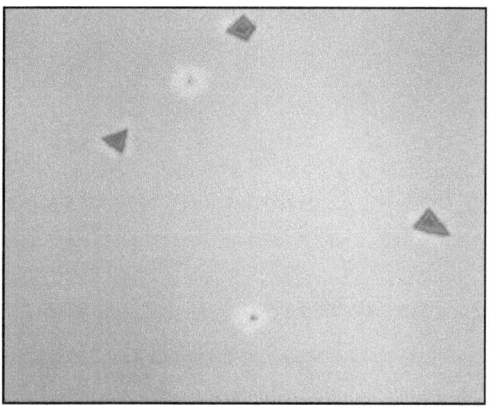

Figure 117 Standard sSOI after etching with the OPE A.
Initial thickness of the sSi layer: ~ 30 nm, thickness after etching: ~ 15 nm,
dip in HF: 5 s, magnification: 1000 times.

Threading dislocations are revealed as tetrahedral etched figures. The strained silicon layer is partially peeled off (SEM image 118). The defect densities are in the same order of magnitude as those of the Secco diluted 1 reference: OPE A ~ 1.1×10^5, Secco dil. ~ 1.5×10^5.

Organic Peracid Etches

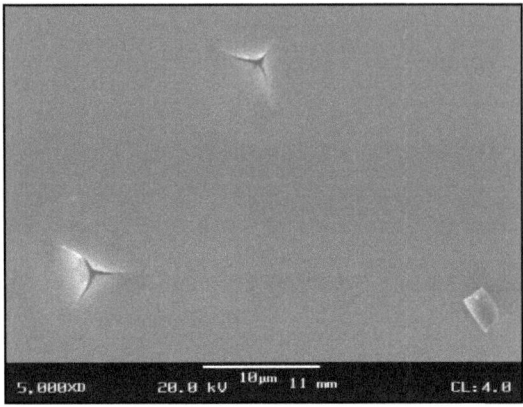

Figure 118: SEM image of standard sSOI after etching with the OPE A.
Initial thickness of the sSi layer: ~ 30 nm, thickness after etching: ~ 15 nm.
Magnification: 5000 times.

The OPE A-D and F were also tested on thin sSOI with an initial thickness of the sSi layer of ~ 15 nm. The Secco diluted 2 solution was used as a reference. All experiments were performed at room temperature (25°C). The removal rates of the different OPE mixtures on sSOI are similar to those determined on SOI material. Six wafer fragments were used for each etching solution and etched down to a thickness of ~ 9 nm. The initial and residual thickness of the sSOI layer was determined using an ellipsometer. The OPE A, B and F revealed threading dislocations in thin sSOI material. They appear as small circular pits (figure 119 – 120). The experimentally determined threading dislocation densities are summarized in table 34. The abovementioned mixtures did not reveal other crystal defects like stacking faults or pile ups, for example. The TD densities are generally lower than those found after etching with the Secco diluted 2 reference (table 34, figure 121).

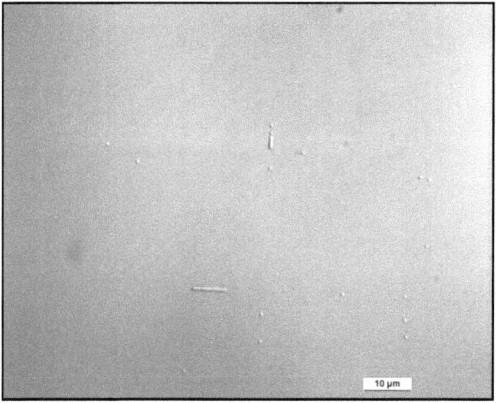

Figure 119: Thin sSOI fragment after etching with the Secco dil. 2 reference. Etching time: 10 s, dip in HF: 10 s, initial thickness: 15 nm, removal: ~ 5.6 nm, magnification: 1000 times. Threading dislocations appear as small circular pits. The thick lines are caused by pile ups.

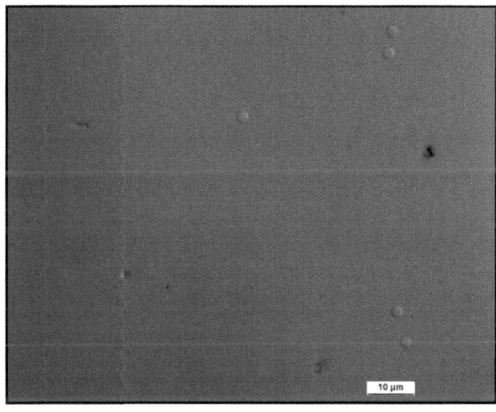

Figure 120: A sSOI sample from the same wafer as above after etching with the OPE F. Etching time: 3 min., dip in HF: 60 s, initial thickness: 15 nm, removal: ~ 5.8 nm, magnification: 1000 times. Threading dislocations appear as circular pits.

Organic Peracid Etches

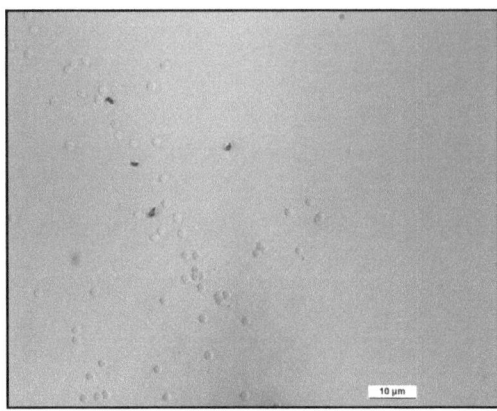

Figure 121: Thin sSOI material after etching with the OPE B.
Etching time: 15 min., initial thickness: 15 nm, removal: ~ 5.8 nm, magnification: 1000 times.
Threading dislocations are also revealed as circular pits.

Etching solution used	Etching time	HF Dip	Thickness after etching [nm]	Defect density/cm^2
Secco diluted 2	**10 s**	**10 s**	**9.4 ± 0.8**	**2.3 x 10^5 ± 1 x 10^5**
OPE A	**10 min.**	**60 s**	**9.2 ± 1.3**	**1.1 x 10^5 ± 4.5 x 10^4**
OPE B	15 min.	-	8.9 ± 2.7	7.5 x 10^4 ± 2 x 10^4
OPE F	3 min.	60 s	8.5 ± 1.2	2.5 x 10^4 ± 1 x 10^4

Table 34: Defect densities found after etching with the OPE A, B and F and the Secco diluted 2 reference.

The TD densities determined for the OPE A correspond very well with those obtained with the Secco diluted 2 reference.

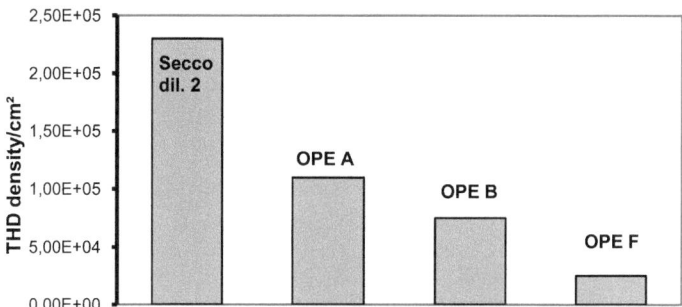

Figure 122: Experimentally determined TD densities.

The Organic Peracid Etches A, B and F were able to reveal threading dislocations on different sSOI materials. Other crystal defects like stacking faults, pile ups or twins, which also appear in the strained silicon film were not delineated.

These experimental results are comparable to those obtained by Taraschi et al. [48, 49].

The TD densities which were obtained after treatment with the OPE A correspond very well with those found after etching with the Secco diluted reference.

6 Characterization of the etching solutions on the basis of their selectivity, activation energy and standard potential

6.1 Introduction

Several chromium based and also chromium free etching solutions that are in use today were described in chapter 4.4 and 4.5. A new class of chromium free etching solutions, the Organic Peracid Etches, was introduced in chapter 5.

Etch formulations can be characterized according to experimental parameters like removal rate, the activation energy for the etching process, standard potential or selectivity. Of these the most important empirical parameter is selectivity.

At a crystal defect or imperfection the removal rate should increase due to a decrease in activation energy for the etching process (chapter 4.3). The increased removal rate leads to the formation of a pit the depth of which can be determined by atomic force microscopy (AFM). The depth of an etch pit should be dependent on how selectively the attack on the crystal defect takes place. The higher the selectivity of the etching solution the deeper the etch pit should be. The Selectivity S expresses the relationship between the removal at the crystal defect and the removal at the perfect crystal.

$$S = \text{selectivity} = \frac{\text{removal}_{\text{(perfect material)}} + \text{depth}_{\text{(etch pit)}}}{\text{removal}_{\text{(perfect material)}}} = \frac{\text{removal}_{\text{(crystal defect)}}}{\text{removal}_{\text{(perfect material)}}}$$

6.2 Experimental determination of the selectivity

Si bulk fragments of approximately 1 square cm with three or four indentations in the centre (see below) were used to determine the selectivity. The physical and chemical properties of the material used are shown in table 35. The indentations cause a strain inside the crystal lattice. After annealing at high temperatures the mechanical stress is reduced by the formation of dislocations. As mentioned in chapter 3, dislocations cannot end inside the crystal lattice. Dislocation loops are formed around the indentations and can be revealed by using a preferential etch.

Characterization of the etching solutions

Material used	Thickness (μm)	Diameter (mm)	Orientation	Dopant/ resistivity (Ωcm)
CZ-bulk Supplier 1	750	300	(100)	p-doped 20-25

Table 35: Physical and chemical properties of the Si-Bulk material used for the experiments

The indentations in the centre of the fragments were generated using an indentor (figure 123). A certain weight was pressed against the crystal surface for ten seconds. It was found that a maximum force of 1N is necessary to produce well-developed indentations (figure 124). At higher loads the fragments start to break.

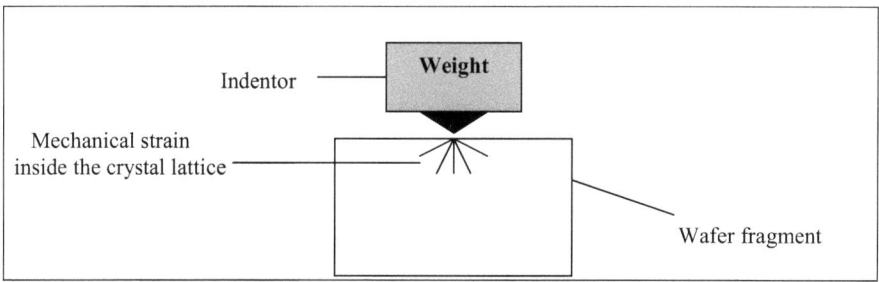

Figure 123: Indentations made by pressing a certain weight against the wafer surface. The indentations cause mechanical damage and strain inside the crystal lattice. A following annealing step leads to the formation of dislocations, primarily in the form of half loops.

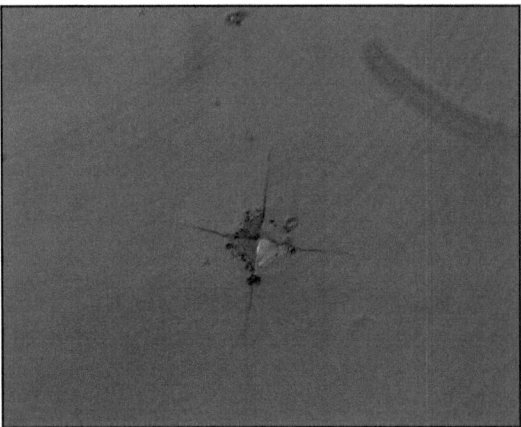

Figure 124: An indentation in the centre of a wafer fragment. Maximum force: 1N, magnification: 1000 times. This fragment was not etched.

The Si bulk fragments with the indentations were heated for 1 minute at 1000°C in an argon atmosphere to generate dislocations. The fragments were then etched with the corresponding solutions (table 36) at 25°C. The temperature was adjusted by using a thermostat. All solutions were able to reveal dislocations which are normally delineated as oval shaped pits (table 36 figures 125-130).

Characterization of the etching solutions

Etching solution used	Capability of solution to reveal dislocations	Characterization of the etched figures	Nature of pits formed
Secco "original" (c Cr^{6+} = 0.1mol/l)	Yes	Oval-shaped pits	double pits
Secco dil. 1 (c Cr^{6+} = 0.03 mol/l)	Yes	Oval-shaped pits	double pits
Secco dil. 2 (c Cr^{6+} = 0.015mol/l)	Yes	Oval shaped pits	double pits
MEMC	Yes	Oval-shaped pits	double pits
Jeita	Yes	Oval-shaped pits	double pits
OPE C	Yes	Oval-shaped pits	double pits
OPE D	Yes	Oval-shaped pits	double pits
OPE F	Yes	polygonal-shaped pits	double pits

Table 36: Characterization of the etch pits found after treatment with different structural etches.

Single pits or a pair of two neighbouring individual etch pits (table 36) were always found. Figures 125 – 130 show SEM images of the etched figures which were found after treatment with the various solutions. The removal on perfect material was calculated (chapter 4.4, 4.5 and 5.2).

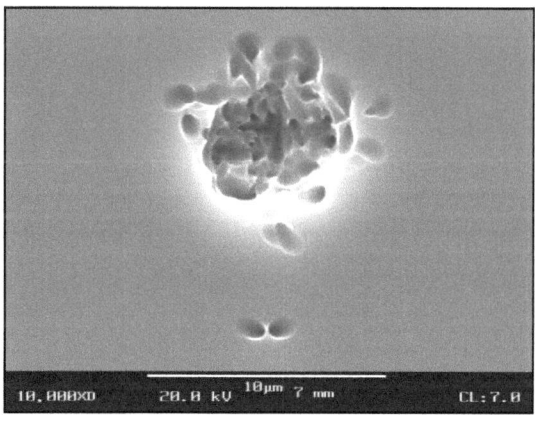

Figure 125: SEM image of oval-shaped etch pits caused by dislocations.
Etching solution used: Original Secco, etching time: 1 min., removal of perfect material: ~ 770 nm, magnification: 10 000 times.

134

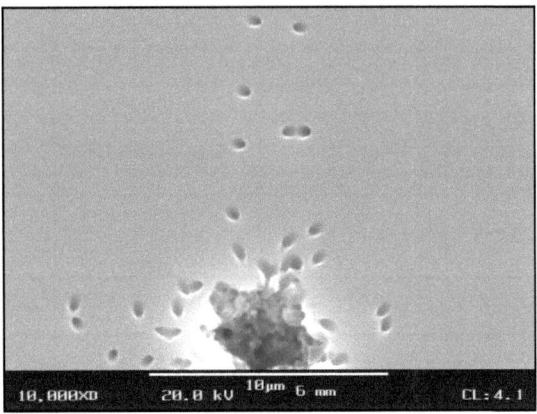

Figure 126: The Secco diluted 1 solution is also able to reveal dislocations. Etching time: 12 min., removal of perfect material: ~ 530 nm, magnification: 10 000 times. The single and double pits which are obtained after etching are also oval-shaped but smaller than those formed with the original Secco.

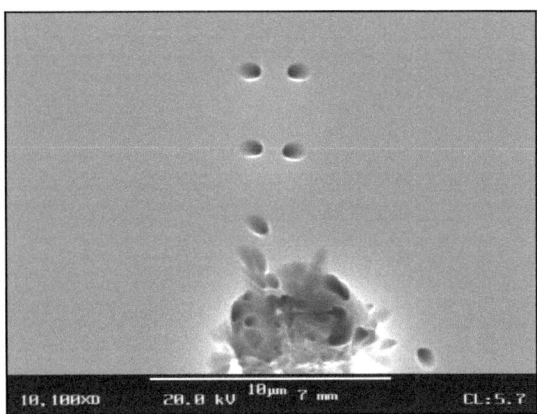

Figure 127: MEMC solution: Pairs of oval-shaped pits found at dislocation half loops. Etching time: 2 min., removal at defect-free site: ~ 940 nm, magnification: 10 000 times.

The Organic Peracid Etches are also able to reveal dislocations as mentioned before. Dislocations are also delineated as oval-shaped pits. The fragments shown below were etched with the OPE C, D and F.

Characterization of the etching solutions

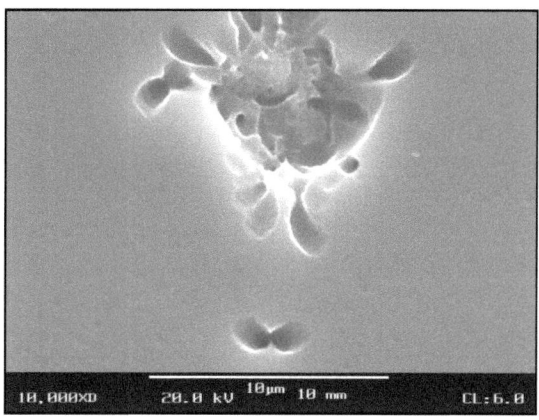

Figure 128: OPE C revealed oval-shaped pits (SEM). Etching time: 12 h, removal at defect-free site: ~ 950 nm, magnification: 10 000 times.

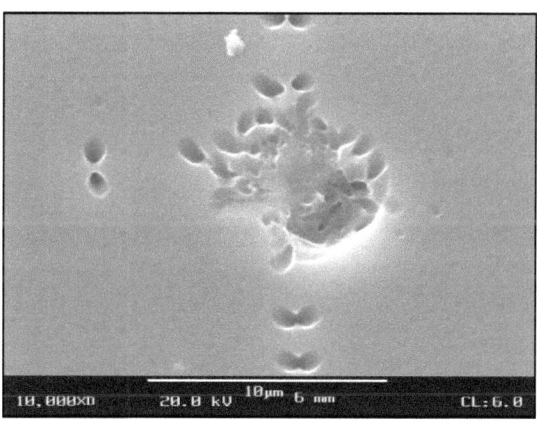

Figure 129: The OPE D has also the capability to reveal crystal defects. Dislocations are also delineated as oval shaped pits. Etching time: 8 h, removal at defect-free site: ~ 850 nm, magnification: 10 000 times.

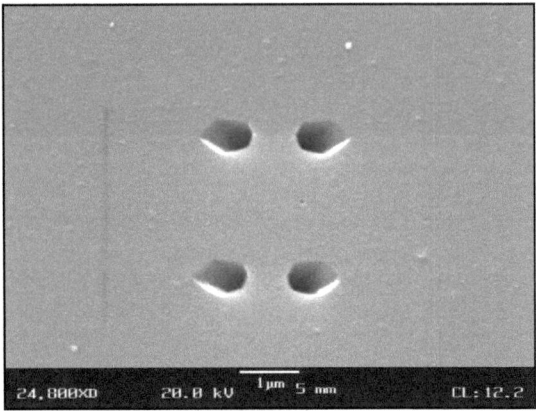

Figure 130: The OPE F produced pairs of polygonal etch pits at dislocation half loops. Etching time: 5 h, removal of perfect material: ~ 520 nm, magnification: ~ 25 000 times.

The pits that are obtained after etching with the various formulations are clearly asymmetric (figure 125 – 130). The depth of the etch pits which are located around the indentations were determined by atomic force microscopy. A tip consisting of silicon with a height of 11 μm and a diameter of approximately 10 nm was always used for AFM investigations. The maximum depth that can be determined by AFM depends on the tip angle and the diameter of the corresponding etch pit (figure 131 equation 31). The tip angle normally lies between 10° and 25°.

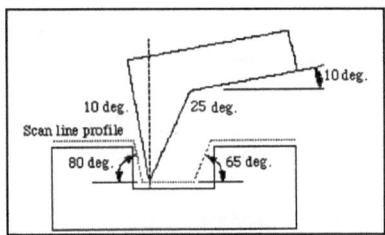

Figure 131: Experimental determination of the depth of a pit by AFM. The maximum depth that can be determined depends on the tip angle α.
Reference: *Olympus technical data sheet.*

As the etching time increases in the case of dislocation half loops the solution preferentially enlarges the pits in a lateral direction and the diameter of the pits increases making them appear shallower than they are. The AFM tip can no longer get to the deepest spot. The determination of selectivity after long etching times is therefore not reliable.

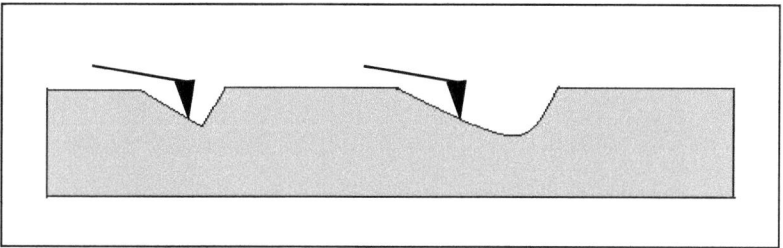

Figure 132: The maximum determinable depth of an etch pit by AFM is dependent on the geometry and depth of the pit.
With short etching times the AFM tip should reach the bottom. In the case of long etching times the pits are laterally enlarged following the dislocation line. In this pit geometry the tip cannot reach the true bottom of the dislocation etch pit resulting in a too low depth.

The maximum depth that can be determined experimentally is given by the following equation:

$$(31)\ \tan \alpha = \frac{a}{b} \quad \rightarrow \quad b = \frac{a}{\tan \alpha}$$

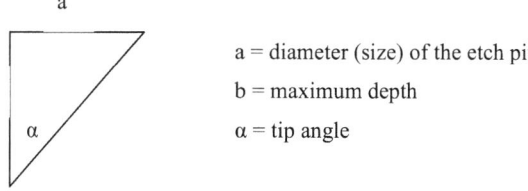

a = diameter (size) of the etch pit
b = maximum depth
α = tip angle

The depth of at least six pits was determined for each solution and etching time. The depth of an etch pit should be dependent on the etching solution used and the etching time. The higher the selectivity of the corresponding solution the deeper the etch pit should be. All AFM measurements were performed in the tapping mode.

Characterization of the etching solutions

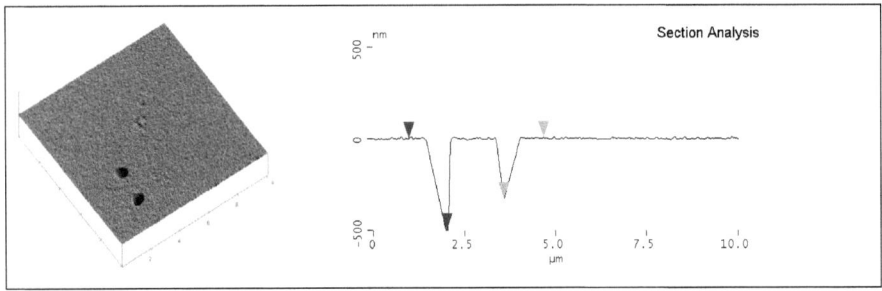

Figure 133: AFM 3D image (scan area 10 µm x 10 µm) and AFM line scan of two single pits found after etching with the Secco solution.
Etching time: 40 s, removal of perfect material: ~ 510 nm, depth of the pits: 523 nm and 330 nm. The etch pits are clearly separated.

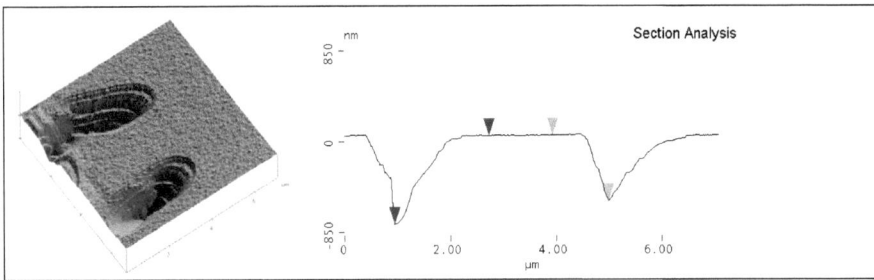

Figure 134: AFM 3D image (scan area 10 µm x 10 µm)
and AFM line scan of two double pits found after etching with the Secco solution. Etching time: 2 min., removal : at defect-free site ~ 1540 nm, depth of the pits: 831 nm and 614 nm. Two previously individual pits have merged to form a double pit.

At a dislocation the removal rate is increased due to an increase in potential energy and, hence, a decrease in activation energy. The increased removal rate can be calculated by dividing the removal at crystal defect by etching time:

$$r_{\text{(dislocation)}} = \frac{\text{removal}_{\text{(perfect material)}} + \text{depth}_{\text{(etch pit)}}}{t_{\text{(etch)}}}$$

The removal on defect free material was calculated and the depth of the pits was determined experimentally. The strain around a dislocation leads to a decrease of activation energy and an increased removal rate. The increase ranges from ~ 40 % to ~ 150 % depending on the etching solution used. The experimental results are summarized in table 37.

Etching solution	Removal rate (nm/min.) perfect material	Removal rate (nm/min.) at crystal defect	Increase (%)
Secco	770 ± 44	1596 ± 193	107
Secco dil. 1	44.4 ± 1.1	109.9 ± 1.4	147
Secco dil. 2	31 ± 1.5	74.8 ± 2.9	141
MEMC	469.2 ± 28	676.6 ± 31	44
Jeita	321 ± 19	470.4 ± 68	47
OPE C	1.35 ± 0.13	2.7 ± 0.36	100
OPE D	1.7 ± 0.12	3.9 ± 0.68	129
OPE F	1.8 ± 0.09	3.8 ± 0.79	111

Table 37: Experimentally determined removal rates of the different etch formulations.

The size of the etch pits is also dependent on the removal which in turn increases with etching time. Short etching times lead to the formation of small and clearly separated individual pits (figure 127, 130 and 133). In the case of dislocation half loops for long etching times, two individual pits can merge to form a double pit (figure 125, 128, 129 and 134). Selectivity has also been found to be dependent on etching time. It decreases with increasing removal (figure 135 and table 38). As mentioned before, in the case of dislocation half loops after prolonged etching times the etching solution preferentially etches in a lateral direction along the dislocation line and the removal rate appears to decrease.

Etching time	Diameter of etch pits	Maximum determinable depth by AFM Tip angle: 10° Tip angle: 25°	Experimentally determined depth	S
15 s	0,4 µm ± 0.08 µm	2.3 µm 0.9 µm	206.7 nm ± 25 nm	1.9 ± 0.13
40 s	0.74 µm ± 0.06 µm	4.2 µm 1.6 µm	426,6 nm ± 97 nm	1.8 ± 0.19
1 min.	1.4 µm ± 0 µm	7.95 µm 3 µm	528.2 nm ± 59 nm	1.7 ± 0.08
2 min.	1.5 µm ± 0.04 µm	8.5 µm 3.2 µm	722.2 nm ± 108.6 nm	1,5 ± 0.07

Table 38: Depth and diameter of the etch pits and selectivity (S) determined for the Secco solution at different etching times.

Characterization of the etching solutions

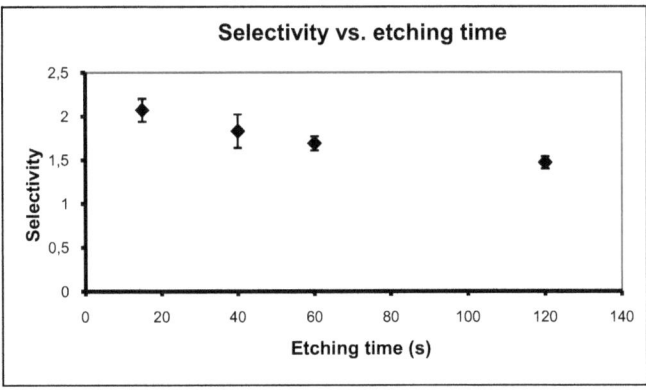

Figure 135: Dependence of the selectivity on etching time determined for the Secco solution. The selectivity apparently decreases with increasing removal.

The highest selectivity can be obtained by extrapolation of the etching time t to 0. Similar experimental results were obtained in case of the Secco diluted 1 and 2 solutions. The selectivity decreases also with increasing etching time (table 39).

Etching solution used	Etching time	Diameter (etch pits)	Maximum determinable depth by AFM Tip angle: 10° Tip angle: 25°	Experimentally determined depth	S
Secco dil. 1	3 min.	0.75 µm ± 0 µm	4.3 µm 1.6 µm	196.5 nm ± 2.5 nm	2.4 ± 0.03
Secco dil. 1	4 min.	0.6 µm ± 0 µm	3.4 µm 1.3 µm	230 nm ± 8.9 nm	2.3 ± 0.05
Secco dil. 1	5 min.	0.4 µm ± 0.16 µm	2.3 µm 0.9 µm	210 nm ± 47.9 nm	1.9 ± 0.2
Secco dil. 1	12 min.	0.8 µm ± 0.03 µm	4.5 µm 1.7 µm	478 nm ± 95.4 nm	1.9 ± 0.18
Secco dil. 2	4 min.	0.3 µm ± 0.03 µm	1.7 µm 0.6 µm	176 nm ± 6.8 nm	2.4 ± 0.05
Secco dil. 2	5 min.	0.5 µm ± 0.06 µm	2.8 µm 1.1 µm	197 nm ± 39.7 nm	2.3 ± 0.26
Secco dil. 2	30. min.	0.9 µm ± 0.18 µm	5.1 µm 1.9 µm	533.9 nm ± 20 nm	1.6 ± 0.02

Table 39: Dependence of the selectivity on etching time determined for the Secco diluted 1 and Secco diluted 2 solution. The selectivity also decreases with increasing removal.

The experimentally determined selectivities for the different Organic Peracid Etches are shown in table 40. Nearly the same selectivities were found for different etching times and OPE mixtures.

Etching solution used	Etching time	Diameter (etch pits)	Maximum determinable depth by AFM Tip angle: 10° Tip angle: 25°	Experimentally determined depth	S
OPE C	3 h	0.7 µm ± 0.14 µm	4 µm 1.5 µm	244.4 nm ± 32.7 nm	2 ± 0.14
OPE C	8 h	1.2 µm ± 0.2 µm	6.8 µm 2.6 µm	648.6 nm ± 81 nm	2 ± 0.13
OPE D	2 h	0.6 µm ± 0.1 µm	3.4 µm 1.3 µm	263 nm ± 46.4 nm	2.3 ± 0.22
OPE D	4 h	1.3 µm ± 0.38 µm	7.4 µm 2.8 µm	479.5 ± 38 nm	2.2 ± 0.1
OPE D	5:43 h	1.9 µm ± 0 µm	10.8 µm 4.1 µm	748.7 nm ± 28.4 nm	2.3 ± 0.05
OPE F	5 h	1.1 µm ± 0 µm	6.3 µm 2.4 µm	602.3 nm ± 126.2 nm	2.1 ± 0.23

Table 40: Experimentally determined selectivities for the OPE C, D and F.

Selectivities were also determined for the Jeita and MEMC solutions. The etching mechanism of silicon in HNO_3/HF mixtures was described in chapter 4.5. Different N(III) species are assumed to be the reactive species which oxidize the silicon. The N(III) species are formed autocatalytically during the etching process. This means that there is an induction period at the beginning of the etching process in which little or no etching takes place. In order to be able to determine their selectivities and compare them with that of the other etching solutions studied, wafer fragments with a size of ~ 1 square cm were first placed in both the Jeita and MEMC solutions for 5 minutes to generate sufficient N(III) species and ensure a homogeneous removal from the onset. They were then removed and the wafer fragments with indentations were etched with the solutions thus prepared.
The experimental results are shown on table 41.

Characterization of the etching solutions

Etching solution used	Etching time	Diameter (etch pits)	Maximum determinable depth by AFM Tip angle: 10° Tip angle: 25°	Experimentally determined depth	S
Jeita	1 min.	0.2 µm ± 0.05 µm	1.1 µm 0.4 µm	149.9 nm ± 21.7 nm	1.5 ± 0.07
MEMC	30 s	0.2 µm ± 0.07 µm	1.1 µm 0.4 µm	69.4 nm ± 13.2 nm	1.3 ± 0.05
MEMC	1 min.	0.5 µm ± 0 µm	0.44 µm 3.85 µm	207.4 nm ± 9.4 nm	1.4 ± 0.02

Table 41: Experimentally determined selectivities for the Jeita and MEMC solution.

The Jeita and the MEMC solutions produce very small and shallow etch pits. Therefore the calculated selectivities are rather low. Both solutions have nearly the same selectivity. This is not surprising because of their similar chemical composition.

The experimental results for all solutions are summarized in table 42.

The highest selectivity, which is that obtained normally after the shortest etching time was always used to compare the formulations with one another.

Etching solution	Removal rate (nm/minute) perfect material	Removal rate (nm/minute) at crystal defect	Increase (%)	S
Secco	770 ± 44	1596 ± 193	107	1.9 ± 0.13
Secco dil. 1	44.4 ± 1.1	109.9 ± 1.4	147	2.4 ± 0.03
Secco dil. 2	31 ± 1.5	74.8 ± 2.9	141	2.4 ± 0.05
MEMC	469.2 ± 28	676.6 ± 31	44	1.4 ± 0.02
Jeita	321 ± 19	470.4 ± 68	47	1.5 ± 0.07
OPE C	1.35 ± 0.13	2.7 ± 0.36	100	2 ± 0.14
OPE D	1.7 ± 0.12	3.9 ± 0.68	129	2.3 ± 0.22
OPE F	1.8 ± 0.09	3.8 ± 0.79	111	2.1 ± 0.23

Table 42: Removal rates determined at defect-free sites and also at dislocations and selectivities of the different etching solutions discussed before.

The Secco diluted solutions and the OPE D show the highest selectivities (table 42 and figure 136). The selectivities of the OPE C, F and the original Secco solution are in the same range. The Jeita and MEMC solution have the lowest selectivities of all tested formulations.

Characterization of the etching solutions

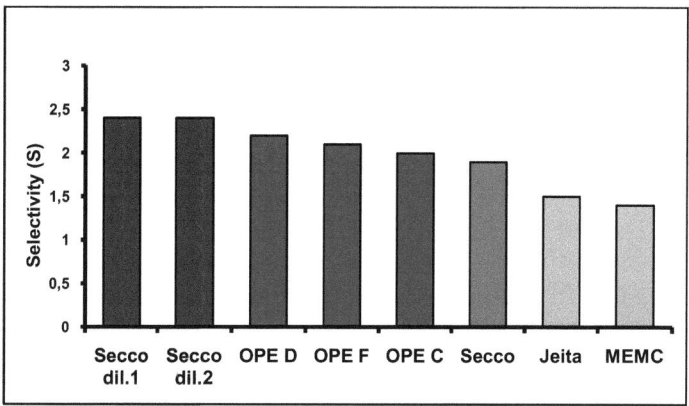

Figure 136: Selectivities of the different etching solutions used.

These experimental results stand in contrast to those of investigations on defect densities. The defect densities found after etching various SOI materials with the OPE C - F were always two to ten times higher than those obtained after etching with the Secco diluted solutions.

Only etching solutions that have a low removal rate are able to delineate even small, point-like defects which is why a higher defect density may be expected of such solutions. Dislocations, however, are relatively large one-dimensional defects and the defect densities obtained thereof should all be of the same magnitude irrespective of the removal rate of the etching solution.

As the defects present in the SOI layer are mainly D-defects and not dislocation loops it is not possible to correlate selectivity with defect density.

6.3 Calculation of the activation energy for the etching process at a crystal imperfection (dislocation)

At a crystal defect the potential energy of the lattice is increased and, hence, the activation energy for the etching process is reduced and therefore the removal rate is increased (table 42). The decreased activation energy can be calculated using the Arrhenius equation:

$\ln r = \ln A - E_a/RT$

$E_a = (\ln A - \ln r)RT$

where Ea = the activation energy of the etching process of silicon, ln r = logarithm of the (increased) removal rate at the defect determined at 25°C and RT = 2.479 kJ/mol (298.15 K). The logarithm of the increased removal rates at the dislocations was calculated from the experimental values shown in table 37 and 43. The pre-exponential factor A was also determined experimentally for each etching solution (chapter 4). The results are summarized in table 43.

Etching solution	Activation energy E_a for the etching process at a defect-free site	Activation energy E_a for the etching process at a dislocation	ΔE_a	ΔE_a (%)
Secco	22.9 kJ/mol	21 kJ/mol	1.9 kJ/mol	8.3
Secco dil. 1	29.3 kJ/mol	27.2 kJ/mol	2.1 kJ/mol	7
Secco dil. 2	36.5 kJ/mol	34.5 kJ/mol	2 kJ/mol	5.5
MEMC	28.6 kJ/mol	28 kJ/mol	0.6 kJ/mol	2
Jeita	36.5 kJ/mol	35.6 kJ/mol	0.9 kJ/mol	2.5
OPE C	48.7 kJ/mol	46.6 kJ/mol	2.1 kJ/mol	4.3
OPE D	46.3 kJ/mol	44.2 kJ/mol	2.1 kJ/mol	4.5
OPE F	42.8 kJ/mol	40.9 kJ/mol	1.9 kJ/mol	4.4
Mean value:				4.8 ± 2

Table 43: Experimentally determined and calculated activation energies for the etching process of silicon.

At a dislocation the activation energy for the etching process is reduced. The calculated values range from 2 % to 8.3 % depending on the etching solution used (table 40).

The mean value lies around 4.8 %.

This means that at a dislocation the inside the crystal lattice leads to an increase of the potential energy and vice versa a decrease of the activation energy of ~ 5 % (table 43) and an increase of the removal rate of ~ 103 %.

6.4 Experimental determination of the standard potentials

The standard potential was also determined experimentally for each etching solution.

A Ag|AgCl electrode was used as the reference electrode and a platinum electrode was used as the indicating electrode. Figure 137 shows the experimental arrangement. The Ag|AgCl electrode is placed inside a vessel containing a 3 M KCl solution. The platinum electrode is plunged into a vessel containing the etching solution. Both electrodes are connected with a potentiometer. The vessels are connected by a diaphragm cell.

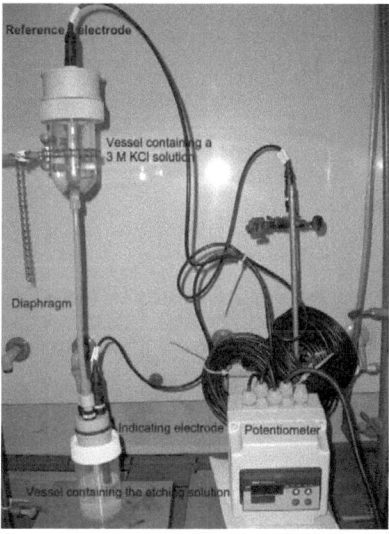

Figure 137: Experimental arrangement used for the determination of the standard potential.

The measured potential is 0.2 V above the standard potential measured with a hydrogen electrode so that 0.2 V was always subtracted from the measured value to calculate the standard potential.

Table 44 shows the standard potentials obtained for the etching solutions studied.

Etching solution used	Standard potential (mV)
Secco	992
Secco diluted 1	828
MEMC	811
Jeita	776
OPE D	506
OPE F	532

Table 44: Experimentally determined standard potentials.

6.5 Comparison of the different etching solutions by their selectivities, activation energies for the etching process and normal potentials

Finally, the different etching solutions were compared in respect of their physical and chemical properties. The most important experimental parameters are summarized in table 45. The activation energies for the etching process of silicon range from ~ 30 kJ/mol to ~ 49 kJ/mol except for the Secco solution which has a significant lower activation energy of approximately 23 kJ/mol. The high activation energies indicate a more reaction controlled etching mechanism of the silicon (chapter 4.2). All solutions which are shown in table 44 are able to delineate different crystal defects (chapter 4.4, 4.5 and chapter 5).

The selectivities range from ~ 1.5 to ~ 2.5. The two Secco diluted solutions and the OPE D and F show the highest selectivities.

Etching solution	Removal rate (nm/minute)	Standard potential (mV)	Activation energy E_a for the etching process (perfect material)	Selectivity
Secco	770	992	22.9 kJ/mol ± 1.3 kJ/mol	1.9
Secco dil. 1	44.4	828	29.3 kJ/mol ± 0.7 kJ/mol	2.4
Secco dil. 2	31	-	36.5 kJ/mol ± 1.5 kJ/mol	2.4
MEMC	469	811	28.6 kJ/mol ± 1.7 kJ/mol	1.4
Jeita	321	776	36.5 kJ/mol ± 2.2 kJ/mol	1.5
OPE C	1.35	-	48.7 kJ/mol ± 4.7 kJ/mol	2
OPE D	1.7	506	46.3 kJ/mol ± 3.3 kJ/mol	2.3
OPE F	1.8	532	42.8 kJ/mol ± 2.1 kJ/mol	2.1

Table 45: A comparison of etching solutions on the basis of various experimentally determined parameters.

The basic idea was to determine different experimental parameters like the removal rate, the activation energy for the etching process of silicon and the standard potential for each etching solution and to compare these parameters with the experimentally determined selectivity. Initially it was assumed that a low removal rate and a high activation energy for the etching process of silicon are synonymous with a high selectivity.

The lower the removal rate, the higher the selectivity of the corresponding etching solution. Therefore the Organic Peracid Etches should have the highest selectivities and the original Secco solution should have the lowest selectivity of all the etching solutions used. This hypothesis was not confirmed.

Characterization of the etching solutions

The dependence of the selectivity on activation energy and removal rate is shown in figures 138, 139 and 140 respectively.

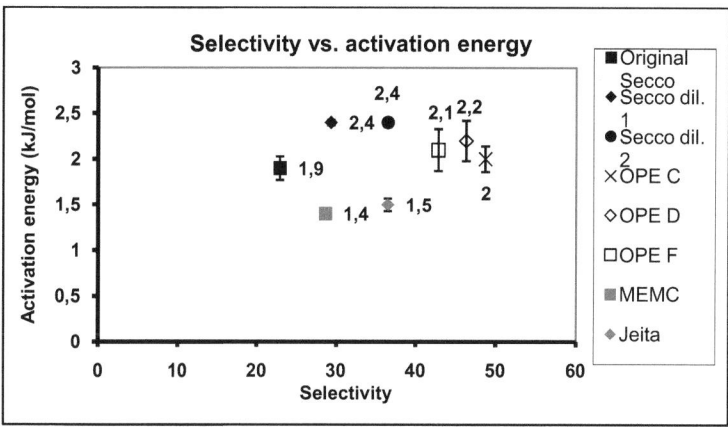

Figure 138: Selectivity as a function of activation energy for the etching process of silicon.

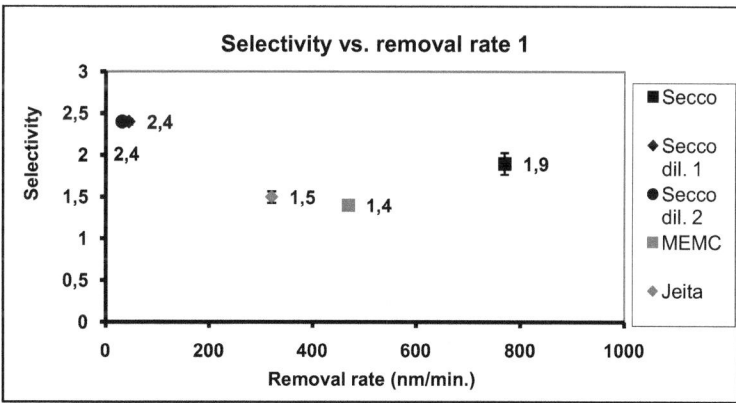

Figure 139: Selectivity as a function of removal rate. Shown for the Original Secco solution, the two different Secco diluted formulations and the MEMC and Jeita solutions.

Characterization of the etching solutions

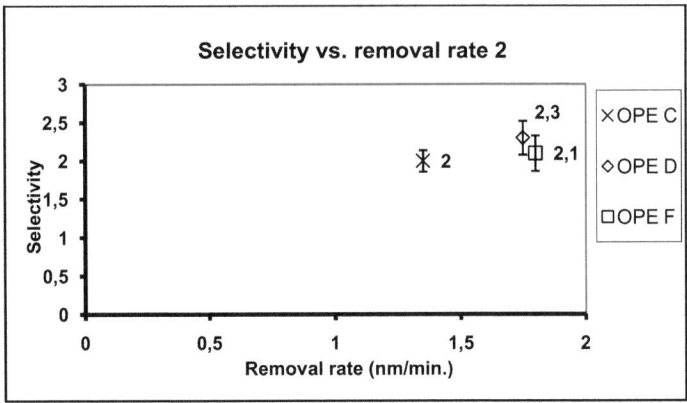

Figure 140: Selectivity dependent on removal rate experimentally determined for the OPE C, D and F.

Obviously there is no correlation between the removal rate or the activation energy for the etching process on the one hand and the selectivity of an etching solution on the other hand. Figure 144 shows the dependence of the selectivity on the standard potential of the etching solution.

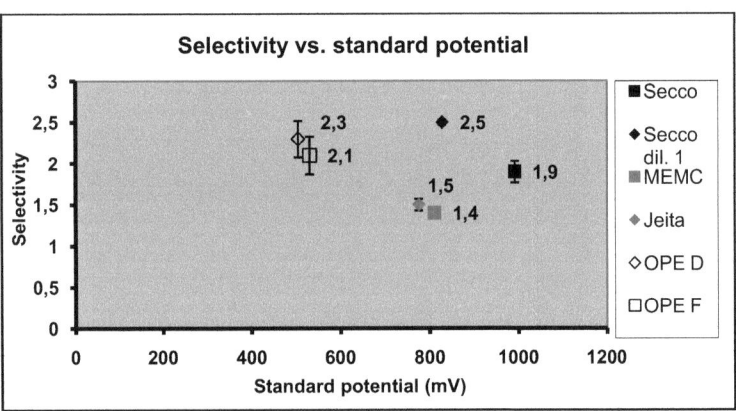

Figure 141: Selectivity as a function of standard potential.

There is also a quite poor correlation between the selectivity of an etching solution and its standard potential. An etching solution with a high standard potential should have strong oxidizing properties Therefore a high normal potential should correspond to a high removal rate and a low activation energy for the etching process and a low selectivity. But the different Secco mixtures which have the highest normal potentials show also the highest selectivities (table 45 and figure 142).

A good correlation was found between the activation energy for the etching process and the standard potential (figure 142).

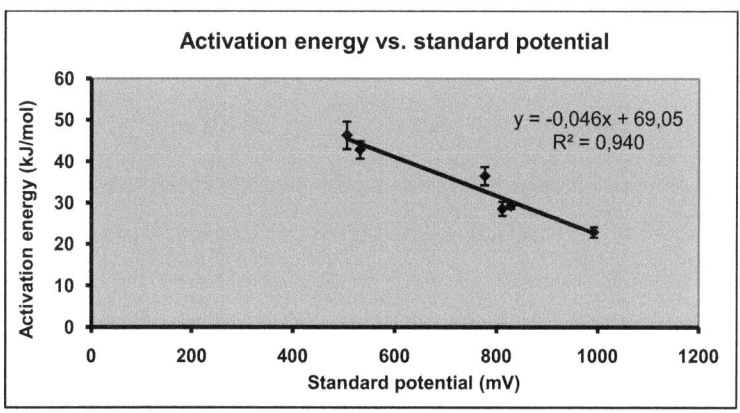

Figure 142: Activation energy for the etching process of silicon as a function of the standard potential of the etching solution.

Unfortunately it is not possible to predict the selectivity of the different etch formulations neither by its composition nor by its chemical or physical properties like removal rate, activation energy for the etching process or standard potential for example. The selectivity has to be determined individually for each etching solution and each kind of crystal defect.

7.1 Summary and Conclusion

Delineation of crystalline defects in semiconducting materials by chemical etching in combination with light optical microscopy is a well established method for quality control. Toxic and carcinogenic chromium(VI) is a component of most structural etching solutions and will be banned in the near future. The most frequently used structural etch is the Secco formulation which is a mixture of $K_2Cr_2O_7$ and hydrofluoric acid. It is able to reveal different kinds of crystal defects in silicon substrates. Diluted versions can be used for the delineation of defects in thin films.

However, most of these etching solutions do not measure up to the demands placed on them today. A new mixture should not only be chromium (VI)-free and have a much lower etch rate, it should also be able to reveal different types of crystal defects with a high selectivity. Furthermore its chemical composition should be simple and the components inexpensive and easy to handle and store.

This study is concerned with a new class of etching solutions called Organic Peracid Etches (OPE) developed to fulfill all the abovementioned requirements.

These solutions consist of hydrogen peroxide, hydrofluoric acid and a short-chain alkanoic acid like acetic- or propanoic acid. When these three components are brought together, a peracid, corresponding to the alkanoic acid employed, is always formed.

The peracid is assumed to be the reactive species which oxidizes the silicon. Solutions containing only hydrogen peroxide and hydrofluoric acid are not able to etch silicon. The formation of the peracid is a typically balanced reaction. The maximum peracid concentration is obtained after 8 to 72 hours, depending on the composition of the etching solution used (chapter 5.4 and 5.6). The removal rates of such mixtures are almost directly proportional to their peracid concentration whereas the hydrofluoric acid concentration has no influence on the removal rate. The oxidation of silicon is the rate-determining step (chapter 4.1 and 4.2).

The thickness of the layer removed with the different OPEs also increases with etching time. The OPE mixtures provide very low removal rates (0.4-1.8 nm/min., Secco diluted: 31 – 44 nm/min., at 25°C) and are able to reveal different kinds of defects like dislocations and vacancy agglomerates (D-defects), swirl-defects (A-defects) or oxidation induced stacking faults (OiSF or OSF).

Due to their low removal rates the Organic Peracid Etches are well suited to the delineation of crystal defects in thin and very thin silicon films. Therefore the OPEs were tested on different

SOI materials produced by the Smart-Cut™ or SIMOX process. A diluted Secco solution was always used as a reference to compare the defect densities (chapter 5.12).

It was found that the defect densities, which were obtained after etching with the different OPE mixtures, were 2 – 10 times higher than those found after etching with the Secco diluted reference. The best match regarding defect densities was obtained with the OPE F. A very good correlation between the defect densities obtained with the OPE F and the Secco diluted reference was found in Smart-Cut™ materials containing oxidation induced stacking faults generated in the SOI fabrication process under non-optimized conditions (figure 146). The defect densities obtained with OPE F, for standard SOI and thin SOI material are in the same range as those found with the Secco diluted reference.

As for SIMOX material the defect densities found after etching with OPE F are significantly higher than those found after etching with the Secco diluted reference (figure 146).

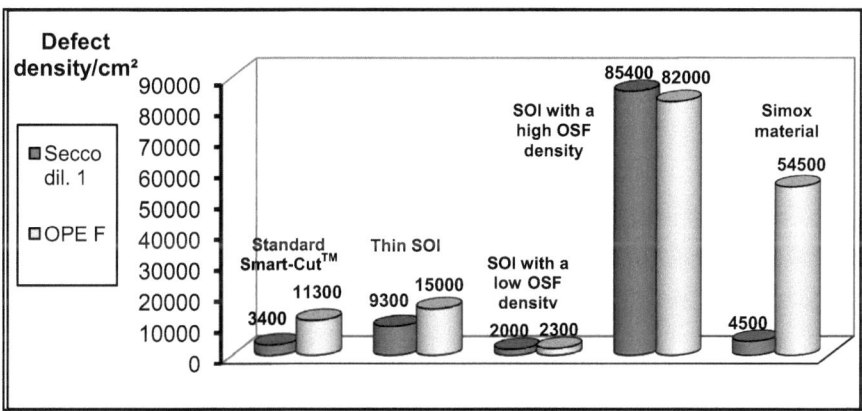

Figure 143: A comparison between the defect densities of the Secco diluted 1 reference and the OPE F

The Organic Peracid Etches produce both etch hillocks and etch pits depending on the hydrofluoric acid content of the solution and the etching time. Detailed SEM investigations revealed a little opening in the centre of the etch hillocks that penetrated the entire silicon layer. It appears that this little hole corresponds to the original defect.

The OPE F is a promising candidate for the delineation and characterization of crystal defects in standard Smart-Cut™ material and thin Smart-Cut™ material.

Oxidation induced stacking faults are revealed with a high selectivity.

The OPE F is thus a suitable substitute for the Secco diluted solutions.

The various OPE mixtures were also tested on sSOI material which consists of a thin strained silicon layer on top of an insulator like silicon dioxide. The OPE A, B and F are able to reveal threading dislocations (TD) in the strained silicon film (chapter 5.11). The TD densities determined for the OPE A correspond very well with those obtained with the Secco diluted reference. The tested OPE mixtures are not able to delineate other crystal defects like stacking faults, pile ups or twins, which also appear in the strained silicon.

Some Organic Peracid Etches were also tested on wafers with an epitaxial silicon layer and on silicon substrates. Epitaxially produced silicon layers are nearly defect-free. Etching times were chosen such that only a part of the epitaxial layer was removed. Nevertheless, after very long etching times (> 16 h) isolated pits were found, with defect densities ranging from $10^4/cm^3$ to $10^6/cm^3$ depending on the etching solution used. No etch pits were found in the remaining epitaxial layer when OPE F was used. Longer etching times appear to favour the formation of artefacts. These artefacts could be caused by the formation of gas bubbles, particles or micro scratches at the crystal surface. The OPE C and D are able to reveal vacancy agglomerates (D-defects) in silicon substrates (see under 5.5, 5.6 and 5.11in chapter 5). Due to their low removal rates and the long etching times which favour the formation of artifacts, these solutions are less suited to the delineation of defects in silicon substrates.

In the second part of this study the different etch formulations have been compared with each other in respect of their physical properties like removal rates, activation energies, standard potentials and selectivities (chapter 6). The selectivity was determined at etch pits caused by dislocations. The depth of the etch pits, determined by atomic force microscopy (AFM), should be dependent on the selectivity of the corresponding etching solution used. The higher the selectivity of the solution the deeper the etch pit should be. It was assumed that a low removal rate and a high activation energy for the etching process should correspond to a high selectivity. However, the experimental results have shown that it is not possible to predict the selectivity of an etching solution from experimental parameters like removal rate or activation energy.

One must bear in mind that selectivity was only determined on one particular type of crystal defect, namely on dislocations. Values for selectivity in the etching solutions can differ for other defect types. Besides the etching solutions used in this study differ considerably from each other in respect of their chemical and physical properties. They can be divided into three completely different etching systems. The original Secco solution and the diluted variations thereof are hydrofluoric acid-dichromate mixtures with the Cr^{6+} species as the oxidizing agent. The Jeita and MEMC solutions contain nitric acid, hydrofluoric acid and, as diluents, acetic acid and water.

Summary

Here the oxidizing agents are various N(III) species which are formed autocatalytically during the etching process. The concentration of acetic acid also plays an important role as it lowers the degree of dissociation of HF and of HNO_3. This has an influence on the pH and the standard potential of the etching solution. The Organic Peracid Etches are mixtures of hydrogen peroxide and a short-chain alkanoic acid like acetic acid. Such systems are strictly speaking not aqueous solutions, the reactive species is the peracid formed. Within each system, however, a certain relationship is perceived between the selectivity of the etching solution on the one hand, and the the activation energy or the removal rate on the other.

The decreased activation energy for the etching process of silicon at a dislocation can be calculated from experimental data by using the Arrhenius equation (chapter 6.3). It was found that the strain inside the crystal lattice caused by a dislocation loop leads to a decrease of the activation energy of ~ 5 % and an increase in the removal rate of ~ 100 %.

8.1 Outlook

The Organic Peracid Etches have been successfully tested on silicon crystals of (100) orientation only. They should, however, also be able to reveal crystal defects in (111) and (110) oriented silicon. They have also been effectively used for the delineation of defects in SOI and should be equally able to reveal defects in germanium on insulator (GOI) which has physical and chemical properties similar to SOI. Removal rates of H_2O_2/HF/HAc mixtures are known to be higher in SiGe alloys than in silicon. They should therefore also be higher in the Ge layer of GOI.

The selectivities of the different etching solutions discussed before were determined at crystal imperfections caused by dislocations. It was found that a dislocation leads to a decrease in activation energy of approximately 5 %. Other crystal imperfections like bulk stacking faults, oxidation induced stacking faults (OSF) or vacancy agglomerates (D-defects) could also cause a strain inside the crystal lattice which should be different from that caused by dislocations. The depth of the corresponding etch pits can also be determined by using atomic force microscopy, and selectivities and activation energies for the etching process calculated in the manner described in chapter 6.

The decrease of the activation energy for the etching process at crystal imperfections can be compared with ab initio or quantum mechanical calculations. Such calculated values could be compared with the experimental data.

7.2 Zusammenfassung

Im Rahmen dieser Arbeit wurde eine neue Klasse von Ätzlösungen zum Nachweis von Kristalldefekten, die sogenannten Organic Peracid Etches, kurz OPE, vorgestellt. Diese Lösungen eignen sich aufgrund ihrer sehr niedrigen Abtragsraten vor allem für die Sichtbarmachung von Kristalldefekten in dünnen Siliziumfilmen, wie sie z.b. in neuartigen Halbleitermaterialien wie Silicon on Insulator (SOI) vorkommen. Die Organic Peracid Etches enthalten eine kurzkettige Alkansäure wie z.b. Essigsäure, Wasserstoffperoxid und Flusssäure. Nach dem Zusammenfügen der Komponenten bildet sich die entsprechende Persäure, z.B. Peressigsäure. Daher rührt auch der Name Organic Peracid Etches, denn die Persäure ist ein stärkeres Oxidationsmittel als Wasserstoffperoxid und vermutlich ist auch die Persäure die reaktive Spezies, welche das Silizium oxidiert. Lösungen, die nur aus H_2O_2 und HF bestehen, ätzen Silizium nur sehr langsam oder gar nicht.

Die Bildung der Persäure ist eine typische Gleichgewichtsreaktion. Je nach Art und Zusammensetzung der Lösung wird die maximale Persäurekonzentration und damit das chemische Gleichgewicht erst nach 8 bis 72 Stunden erreicht (Kapitel 5.4 und 5.6).

Man sollte annehmen, dass die Abtragsrate, d.h. die Ätzgeschwindigkeit von Silizium sowohl von der HF-Konzentration als auch von der Persäurekonzentration abhängt. Dies konnte experimentell jedoch nicht bestätigt werden. Die Abtragsrate hängt nicht von der Flusssäurekonzentration sondern nur von dem Persäuregehalt der Lösung ab. Die Abtragsrate ist nahezu linear von der Persäurekonzentration abhängig (Kapitel 5.3 und 5.4). Die OPE-Lösungen besitzen, wie bereits erwähnt, sehr niedrige Abtragsraten (0,4 nm/Minute – 2 nm/Minute, Secco diluted: 31 – 44 nm/Minute). Sie sind allerdings in der Lage verschiedene Arten von Kristalldefekten wie z.B. Versetzungen, Swirl-Defekte (A-Defekte), Leerstellenagglomerate (D-Defekte) oder auch durch Oxidation induzierte Stapelfehler (OiSF bzw. OSF) sichtbar zu machen. Aufgrund ihrer sehr niedrigen Abtragsraten wurden sie ausgiebig an verschiedenen SOI-Materialien getestet (5.12). Die selben Materialien wurden auch mit einer chromhaltigen Referenzätzlösung behandelt um die Defektdichten vergleichen zu können. Die Defektdichten, welche man nach dem Ätzen mit den OPE-Lösungen findet, sind im Allgemeinen 2 bis 10 mal höher im Vergleich zur Secco diluted Referenz.

Die Organic Peracid Etches weisen eine hohe Sensitivität gegenüber Versetzungen auf. Sie sind deswegen auch in der Lage durch Oxidation induzierte Stapelfehler (OSF), welche teilweise prozessbedingt in SOI-Filmen vorkommen, nach dem Ätzen sichtbar zu machen. Zwei

Zusammenfassung

verschiedene SOI-Substrate, welche nach dem Smart-Cut^{TM}-Verfahren hergestellt wurden und eine hohe respektive eine niedrige OSF-Dichte aufweisen, wurden mit den verschiedenen OPE Mixturen bzw. der chromhaltigen Referenzlösung behandelt. Die Defektdichten, die man nach dem Ätzen mit der OPE F-Lösung bzw. der Referenz erhält, sind nahezu identisch.

Auch Standard Smart-Cut^{TM} SOI sowie sehr dünnes SOI (Schichtdicke des Siliziumfilms (< 30 nm) wurde mit der OPE F-Lösung behandelt. Die Defektdichten stimmen allerdings nicht mehr so gut mit der Referenz überein. Allerdings liegen sie noch in der selben Größenordnung. Die Defektdichten, welche man nach dem Ätzen von SIMOX-Material findet, liegen im Falle der OPE F eine Zehnerpotenz über denen der Referenzlösung.

Die OPE F-Lösung eignet sich sehr gut für die Visualisierung von Kristalldefekten in verschiedenen SOI-Substraten, welche nach dem Smart-Cut^{TM}-Verfahren hergestellt wurden. Sie könnte die chromhaltigen verdünnten Secco-Lösungen, die heute noch routinemäßig eingesetzt werden, ersetzen.

Die OPE-Lösungen wurden auch an verschiedenen sSOI –Materialien getestet. Strained Silicon on Insulator (sSOI) besteht aus einem dünnen Film von gespannten Silizium oberhalb eines Isolators wie Siliziumdioxid. Dieser dünne Film enthält Fadenversetzungen, sogenannte Threading Dislocations (TD). Die OPE A, B und F sind in der Lage solche Fadenversetzungen sichtbar zu machen (Kapitel 5.11). Die Defektdichten, die man nach dem Ätzen mit der OPE A erhält, stimmen sehr gut mit der Secco diluted-Referenz überein. Allerdings sind die OPE-Lösungen nicht in der Lage andere Kristalldefekte wie, Stacking Faults, Pile Ups or Twins, die ebenfalls typischerweise in dem gespannten Siliziumfilm vorkommen, sichtbar zu machen. Einige OPE Mixturen sind auch in der Lage Leerstellenagglomerate, sogenannte D-Defekte oder COPs, in Silizium Substraten zu visualisieren (Kapitel 5.5, 5.6 und 5.11). Allerdings zeigte sich bei Versuchen an Epi-Wafern, dass sehr lange Ätzzeiten die Bildung von Artefakten und eine deutliche Aufrauung der Oberfläche bewirken. Die Bildung von Artefakten könnte durch kleinste Beschädigungen bzw. Risse auf der Waferoberfläche, durch Gasblasen, die während dem Ätzen entstehen oder auch durch eine Abscheidung von Reaktionsprodukten verursacht werden. Aufgrund der sehr langen Ätzzeiten und der damit verbundenen Anfälligkeit bezüglich der Bildung von Artefakten, sind die OPE-Lösungen nicht oder nur bedingt für die Sichtbarmachung von Kristalldefekten in Siliziumsubstraten geeignet.

Die Prozesse, die während des Ätzens von Silizium ablaufen, sind in den meisten Fällen nur unvollständig oder gar nicht verstanden. Das Auffinden bzw. Entwickeln neuer Defektätzlösungen ist daher eher von empirischer Natur.

Zusammenfassung

Die verschiedenen chromhaltigen und chromfreien Ätzlösungen die heute verwendet werden, sowie die Organic Peracid Etches wurden hinsichtlich ihrer physikalischen und chemischen Eigenschaften hin untersucht und verglichen. Dabei wurde auch die experimentell zugängliche Messgröße Selektivität (S) eingeführt, um die verschiedenen Ätzlösungen bezüglich ihrer Fähigkeiten Defekte sichtbar zu machen, besser vergleichen zu können.

An einem Kristalldefekt ist die potenzielle Energie erhöht und damit die Aktivierungsenergie für den Ätzprozess erniedrigt, was zur Ausbildung einer Ätzgrube führt. Je selektiver bzw. empfindlicher der Angriff einer Ätzlösung an Kristalldefekten erfolgt, desto tiefer bzw. größer sollte die Ätzgrube sein. Die Tiefe bzw. der Durchmesser kann mit Hilfe der Rasterkraftmikroskopie (AFM) näherungsweise bestimmt werden.

Als Selektivität (S) wurde dabei das Verhältnis zwischen der erhöhten Abtragsrate am Defekt und der Abtragsrate am perfekten, nicht gestörten Siliziumkristall, definiert. Um die Selektivität bestimmen zu können, wurden Bruchstücke aus Siliziumsubstraten mit Härteeindrücken versehen und anschließend unter Argonatmosphäre bei 1000°C erhitzt. Dadurch entstehen rings um den Eindruck Versetzungsschleifen, welche durch Ätzen sichtbar gemacht werden können.

Die so präparierten Bruchstücke wurden mit den verschiedenen Lösungen unterschiedlich lang geätzt und die Tiefe der Ätzgruben mittels AFM bestimmt. Die verdünnten Secco Lösungen weisen die höchsten und die Jeita bzw. MEMC Lösungen die niedrigsten Selektivitäten auf. Die Annahme, dass die Lösungen mit der niedrigsten Abtragsrate auch am selektivsten angreifen, konnte hiermit nur bedingt bestätigt werden.

Die Selektivität wurde auch mit anderen experimentell leicht zugänglichen Daten wie Aktivierungsenergie für den Ätzprozess von Silizium oder dem Normalpotential einer Lösung verglichen. Hierbei stellte sich heraus, dass kein allgemein gültiger Zusammenhang zwischen der Selektivität einer Ätzlösung auf der einen Seite und ihren physikochemischen Eigenschaften wie z.B. der Abtragsrate, Aktivierungsenergie oder dem Normalpotential auf der anderen Seite besteht.

Man kann leider nicht anhand experimentell leicht bestimmbarer Parameter Rückschlüsse auf die Selektivität einer Defektätzlösung ziehen. Allerdings muss man berücksichtigen, dass die Selektivität nur an einer bestimmten Art von Kristalldefekten, nämlich den Versetzungen bestimmt wurde. Die Ätzlösungen können bei anderen Defekttypen durchaus unterschiedliche Selektivitäten aufweisen. Außerdem unterscheiden sich die untersuchten Lösungen gravierend hinsichtlich ihrer chemischen und physikalischen Eigenschaften. Sie lassen sich in drei völlig unterschiedliche Ätzsysteme einteilen. Bei der original Secco-Lösung und deren verdünnten Varianten handelt es um Flusssäure/Dichromat-Mischungen. Das Oxidationsmittel ist hier eine

Cr^{6+}-Spezies. Die Jeita- und MEMC-Lösungen enthalten Salpetersäure, Flusssäure sowie Essigsäure und Wasser als Diluenten. Das Oxidationsmittel sind hier diverse N(III)-Spezies, die während dem Ätzvorgang autokatalytisch gebildet werden. Auch die Essigsäurekonzentration spielt eine wichtige Rolle, da sie den Dissoziationsgrad der HF bzw. der HNO_3 je nach Konzentration erniedrigt und somit auch den pH-Wert bzw. das Normalpotential der Ätzlösung beeinflusst. Die Organic Peracid Etches sind Mischungen aus Wasserstoffperoxid und einer kurzkettigen Alkansäure wie z.B. Essigsäure. Bei diesen Systemen handelt es sich strenggenommen nicht mehr um wässrige Lösungen, die reaktive Spezies ist hierbei die gebildete Persäure. Innerhalb der einzelnen Systeme kann man allerdings durchaus gewisse Tendenzen bezüglich der Selektivität einer Ätzlösung auf der einen, und der Aktivierungsenergie, der Abtragsrate bzw. dem Normalpotential auf der anderen Seite erkennen.

Zuletzt wurde die erniedrigte Aktivierungsenergie für den Ätzprozess an einer Versetzung mit Hilfe der experimentell gewonnenen Daten berechnet. Es zeigt sich, dass eine Versetzung zu einer Absenkung der Aktivierungsenergie von ca. 5% führt, was sich in einer um rund 100% erhöhten Abtragsrate niederschlägt.

8.2 Ausblick

Die OPE-Lösungen wurden erfolgreich an Siliziumkristallen mit einer 100 Orientierung getestet. Sie sollten daher auch in der Lage sein, Defekte in 111 oder 110 orientiertem Silizium sichtbar zu machen.

Wegen ihrer sehr niedrigen Abtragsraten eignen sich die OPE-Lösungen zur Visualisierung von Kristalldefekten in dünnen bzw. sehr dünnen Siliziumfilmen wie z.B. im SOI-Material.

Aufgrund ähnlicher chemischen und physikalischer Eigenschaften sollten die Organic Peracid Etches auch in der Lage sein, Kristalldefkte in dünnen Germaniumfilmen, wie sie z.b. in Germanium On Insulator (GOI) vorkommen, sichtbar zu machen. Die Abtragsrate sollte auf den Germaniumfilmen deutlich erhöht sein. Es ist nämlich bekannt, dass Essigsäure/HF/H_2O_2-Mischungen Si/Ge-Legierungen mit einer bis zu 30 mal höheren Geschwindigkeit als Silizium ätzen. Die Selektivitäten der verschiedenen Ätzlösungen wurden an Versetzungen bestimmt. Prinzipiell verursachen auch Stapelfehler im Siliziumkristall bzw. auch durch Sauerstoff induzierte Stapelfehler (OSF) eine Störung bzw. Verspannung des Gitters. Stapelfehler führen daher ebenfalls zur Ausbildung von Ätzgruben. Auch Leerstellenagglomerate (D-Defekte oder COPs) könnten Spannungen im Siliziumkristall verursachen, was letzten Endes zu der Bildung von Ätzgruben führt.

Die Tiefe der entsprechenden Gruben und damit auch die Selektivität könnte, wie zuvor schon anhand der Versetzungen diskutiert wurde, mittels AFM bestimmt werden. Die verschiedenen Kristalldefekte sollten das Gitter unterschiedlich stark verspannen. Man sollte daher, abhängig von der Art der Kristalldefekte bzw. der verwendeten Ätzlösung, auch unterschiedliche Selektivitäten erwarten.

9 Appendix

9.1 Bibliography

[1] Integrated circuit (electronics) Britannica Online Encyclopedia
http://www.britannica.com/Ebchecked/topic/289645/integrated-circuit

[2] Uniprotokolle, Die Adresse für Ausbildung, Studium und Beruf
http://www.uni-protokolle.de/Lexikon/Integrierter_Schaltkreis.html

[3] Fumio Shimura
Semiconductor Silicon Crystal Technology
First Edition (1989), Academic Press, Inc

[4] Chromium free etching solution for si-substrates and uses therefore
Patent published by the United States Patent and Trademark Office (USPTO)
http://www.freshpatents.com/Chromium-free-etching-solution-for-si-substrates

[5] Health Protection Agency
HPA Compendium of Chemical Hazards: Chromium
Version 1 (2007), p 22 – 35

[6] G. Wenski, G. Hohl, P. Storck, I. Crössmann
Die Herstellung von Reinstsiliciumscheiben
Chemie Unserer Zeit (37), 2003, p 198 – 208

[7] J. Evers, P. Klüfers, R. Staudigl und P. Stallhofer
Czochralskis schöpferischer Fehlgriff: ein Meilenstein auf dem Weg in die Gigabit-Ära
Angew. Chem. (115), 2003, p 5862 – 5877

[8] Silicon Valley Microelectronics, Inc
http://www.svmi.com/siliconwafers

[9] Karen A. Reinhardt, Werner Kern
Handbook of Silicon Wafer Cleaning Technology
Second Edition (2002), William Andrew

[10] D. Possner, B.O. Kolbesen, H. Cerva, V. Klüppel
Organic Peracid Etches: a new class of chromium free etch solutions for the delineation of defects in different semiconducting materials
ECS Transactions Vol. 10 (1), 2007, p 21 – 31

[11] C. Jaussaud, J. Margail, J. M. Lamure, M. Bruel
Simox technology: From basic research to industrial developments
Radiation Effects and Defects in Solids 127 (3), 1994, p 319 – 326

[12] C. Mazure, and A.J. Auberton-Herve
    SOI: Materials to Systems
    Proceedings of the 35th European Solid State Device Research Conference (IEEE), 2005, p 29

[13] C. Mazure and G.K. Keller
    Advanced Electronic Substrates for the Nanotechnology Era
    The Electrochemical Society Interface (Winter 2006), p 33 – 40

[14] S .H. Christiansen, M. Reiche, R. Singh, I. Radu and R. Scholz
    Ultrathin strained silicon layers on insulators (sSOI) by wafer bonding
    ATMI Epitaxial Services Mesa, AZ, USA

[15] Bich-Yen Nguyen et al.
    New Ideas for New Materials: Advanced Substrates and Devices for Nanoscale CMOS
    Yield Management Solutions (Summer 2004)
    http://www.kla-tencor.com/magazine

[16] R. Falster and V.V. Voronkov
    Intrinsic Point Defects and Their Control in Silicon Crystal Growth and Wafer Processing
    MRS Bulletin (June 2006), p 28

[17] BMD (Bulk Micro Defects)
    http://www.semilab.hu/bmdsemiapp.html

[18] H Rauh
    Wacker´s Atlas For Characterization Of Defects In Silicon
    Wacker Siltronic AG, Burghausen

[19] V. V. Voronkov
    Grown-in defects in silicon produced by agglomeration of vacancies and self-interstitials
    Journal of Crystal Growth (310), 2008, p 1307

[20] A. J .R. de Kock
    Philips res. Repts. Suppl. 1 (1973)

[21] H. Föll, B.O. Kolbesen
    Formation and Nature of Swirl Defects in Silicon
    Applied Physics (8), 1975, p 319 – 331

[22] Takemi Ueki
    Octahedral void defects observed in the bulk of Czochralski silicon
    Appl. Phys. Lett. 70 (10), 1997, p 1248 – 1250

[23] V. V. Voronkov, R. Falster
    Vacancy-type microdefect formation in Czochralski silicon
    Journal of Crystal Growth (194), 1998, p 76 – 78

[24] V. V. Voronkov, R. Falster
Grown-in microdefects, residual vacancies and oxygen precipitation bands in Czochralski silicon
Journal of Crystal Growth (2044), 1999, p 462 – 463

[25] R. C. Newman
Oxygen diffusion and precipitation in Czochralski silicon
J. Phys. Condens. Matter (12), 2000, p 335 – 337

[26] G. Kissinger and J. Dabrowski
Oxide Precipitation Via Coherent Seed-Oxide Phases
Journal of the Electrochemical Society 155 (6), 2008, p 448 – 454

[27] K. Sueko, N. Ikeda, T. Yamamoto and S. Kobayashi
Morphology and growth process of thermally induced oxide precipitates in Czochralski silicon
J. Appl. Phys. 74 (9), 1993, p 5437

[28] T. Sinno, R :A. Brown, W .v. Ammon and E. Dornberger
Point Defect Dynamics and the Oxidation-induced Stacking –Fault Ring in In Czochralski-Grown Silicon Crystals
Journal of the Electrochemical Society Vol. 145 (1), 1998, p 302 – 303

[29] K. Marsden, S. Sadamitsu, M. Hourai, S. Sumita and T. Shigematsu
Observation of Ring-OSF Nuclei in CZ-Si Using Short-Time Annealing and Infrared Light Scattering Tomography
Journal of the Electrochemical Society Vol. 142 (3), 1995, p 996 – 998

[30] H. Yamagishi, I. Fusegawa, N. Fujimaki and M. Katayama
Recognition of D defects in silicon single crystals by preferential etching and effect on gate oxide integrity
Semicond. Sci. Technol. (7), 1992, p 135 – 140

[31] T. Müller, A. Zeller and M. Stallmann
Den Cops Auf der Spur
Wacker World Wide 1/2004
Wacker Siltronic AG, Burghausen

[32] A.F. Bogenschütz
Ätzpraxis für Halbleiter
Hanser Verlag München, 1967

[33] F. Secco d´ Aragona
Dislocation Etch for (100) Planes in Silicon
Journal of the Electrochemical Society: Solid-State Science and Technology Vol. 119 (7), 1972, p 948 – 951

[34] B.O. Kolbesen, J. Mähliß and D. Possner
Chemical Etching Processes for Delineation of Crystalline Defects in Silicon Substrates and Thin Films
Freiberger Silicon Days (June, 2007)

[35] J. E. A. M. van den Meerakker and J. H. C. van Vegchel
Silicon Etching in CrO_3-HF Solutions
Journal of the Electrochemical Society Vol. 136 (7), 1989, p 1949 - 1953

[36] Margaret Wright Jenkins
A New Preferential Etch for Defects in Silicon Crystals
Journal of the Electrochemical Society: Solid-State Science and Technology Vol 124 (5), 1977, p 757 – 761

[37] J. Maehliss, A. Abbadie and B.O. Kolbesen
A Chromium-free Defect Etching Solution for Application on SOI
ECS Transactions Vol. 6 (4), 2007, p 271 – 277

[38] M. Steinert, J. Acker, S. Oswald and K. Wetzig
Study on the Mechanism of Silicon Etching in HNO_3-Rich HF/HNO_3 Mixtures
J. Phys. Chem. C (111), 2007, p 2133 – 2140

[39] H. Robbins and B. Schwartz
Chemical Etching of silicon: II The System HF, HNO_3, H_2O and $HC_2H_3O_2$
Journal of the Electrochemical Society Vol 107 (2), 1961, p 108 – 111

[40] M. Kelly, J. Chun, and A. Borcasly
High efficiency chemical etchant for the formation of luminescent porous silicon
Appl. Phys. Letters 64 (13), 1994, p 1693 - 1695

[41] M. Steinert, J. Acker, S. Oswald and K. Wetzig
Reactive Species Generated during Wet Chemical Etching of silicon in HF/HNO_3 Mixtures
J. Phys. Chem.B, 2006, p 1 – 6

[42] Daniel Swern
Organic Peracids
J. Am. Chem. Soc. (70), 1948, p 2 – 11

[43] J. D.´Ans u. W. Frey
Untersuchungen über die Bildung v. Persäuren aus organischen Säuren und Hydroperoxid
Z. anorg. Chem. (Bd 84), 1913, p 145 – 164

[44] M. Guder
Diploma Thesis
J. W. G. University, Frankfurt am Main, 2005

[45] W. Wijaranakula
Characterization of Crystal Originated Defects in Czochralski Silicon Using Nonagitated Secco Etching
Journal of the Electrochemical Society Vol. 141 (11), 1994, p 3273 – 3277

[46] J. Stoemenos
Structural defects in SIMOX
Nuclear Instruments and Methods in Physics Research B (112), 1996, p 206 – 213

[47] Markus Sulzberger
Wachstum von dreidimensionalen Germaniuminseln auf verspannten und unverspannten Silizium (001)-Oberflächen
Phd Thesis
RWTH Aachen, 2003

[48] G. Taraschi, A. J. Pitera, L. M. Mc Gill, Z. Cheng, M. L. Lee, T. A. Langdo and E. A. Fitzgerald
Ultrathin Strained Sion-Insulator and SiGe-on-Insulator Created using Low Temperature Wafer Bonding and Metastable Stop Layers
Journal of the Electrochemical Society Vol. 151 (1), 2004, p 47 – 56

[49] G. Taraschi, A. J. Pitera and E. A. Fitzgerald
Strained Si, SiGe, and Ge on-insulator: review of wafer bonding fabrication techniques
Solid-State Electronics (48), 2004, p 1297 – 1305

9.2 Instruments used

Light microscopy: Reichert Univar

Digital camera: Leica DFC 280

Software: Leica IM 50 Image Manager

SEM: Atomica/Amray 1920 Eco Environment controlled SEM

AFM: Digital Instruments Nano Scope III a

AFM-tips: Olympus Standard Silicon probe OMCL-AC160TS-C2

Indentor: Mikro-Härteprüfer VMHT-MOT Leica UHL

Profilometer: Tencor Instruments alpha-step 200

Elipsometer: Plasmos SD 2000

Thermostat: VWR Kältethemostat 1160 S

Potentiometer: Jumo dTRANS pH 01
Model: 202530/10-888000-23003000000

Indicating electrode: Jumo Redox-Metallelektrode
Model: 201080/13-85-22-22-120

Reference electrode: Jumo Bezugselektrode (Ag/AgCl)
Model: 201080/11-89-04-07-22-120

KCl storage vessel: Jumo
Model: 20 2810

Diaphragm: Jumo
Model: 201080/15-87-04-22-120

9.3 Chemicals used

H_2O_2 (30 %): Gigabit Ashland Chemical
LOT: 81100451

H_2O_2 (50 %): PA Acros Organics (Stabilizer < 200 ppm)
LOT: A0256632

HF (50 %): VLSI Selectipur BASF Feinchemikalien
LOT: 107 550 2400

Acetic acid (100 %): PA Merck KGaA
LOT: K 39595863

Propanoic acid (99 %): PA Acros Organics
LOT: A0262843

Butyric acid (99 %): PA Acros Organics
LOT: A0223367

HNO_3 (69 %): Selectipur Merck KGaA
LOT: K34387745

Potassium dichromate: PA Sigma-Aldrich
LOT: 62420

Apiezon Wax W 100 Roth

Toluene: PA Sigma-Aldrich
LOT: 0891961

H_3PO_4 (85 %): PA Grüssing

Potassium phosphate (99 %): PA Acros Organics
LOT: A0252177
Potassium iodide briquettes (99 %): PA Acros Organics
LOT: A020063501

H_2SO_4 (96 %): VLSI Selectipur Merck KGaA
LOT: K34775009

Catalase from Micrococcus lysodeikticus: Fluka
LOT and filling code: 121995022606115

Die VDM Verlagsservicegesellschaft sucht für wissenschaftliche Verlage abgeschlossene und herausragende

Dissertationen, Habilitationen, Diplomarbeiten, Master Theses, Magisterarbeiten usw.

für die kostenlose Publikation als Fachbuch.

Sie verfügen über eine Arbeit, die hohen inhaltlichen und formalen Ansprüchen genügt, und haben Interesse an einer honorarvergüteten Publikation?

Dann senden Sie bitte erste Informationen über sich und Ihre Arbeit per Email an *info@vdm-vsg.de*.

Sie erhalten kurzfristig unser Feedback!

VDM Verlagsservicegesellschaft mbH
Dudweiler Landstr. 99   Telefon +49 681 3720 174
D - 66123 Saarbrücken   Fax   +49 681 3720 1749

www.vdm-vsg.de

Die VDM Verlagsservicegesellschaft mbH vertritt

Printed by Books on Demand GmbH, Norderstedt / Germany